Renewable and Sustainable Composites

Edited by António B. Pereira and Fábio A. O. Fernandes

Published in London, United Kingdom

IntechOpen

Supporting open minds since 2005

Renewable and Sustainable Composites
http://dx.doi.org/10.5772/intechopen.77412
Edited by António B. Pereira and Fábio A. O. Fernandes

Contributors
Oscar Marcelo Suarez, John Lopez, Zuleika Oquendo, Hammat Valiev, Alexander Vlasov, Yeeli Kelvii Kwok, Seyyed Alireza Hashemi, Seyyed Mojtaba Mousavi, Aziz Babapoor, Yousef Amrollahi, Shesan Owonubi, Stephen Agwuncha, Chioma Anusionwu, Neerish Revaprasadu, Rotimi Sadiku, António Bastos Pereira, Fabio A.O. Fernandes

First published in London, United Kingdom, 2019 by IntechOpen
IntechOpen is the global imprint of INTECHOPEN LIMITED, registered in England and Wales, registration number: 11086078, 7th floor, 10 Lower Thames Street, London, EC3R 6AF, United Kingdom
Printed in Croatia

British Library Cataloguing-in-Publication Data
A catalogue record for this book is available from the British Library

Additional hard and PDF copies can be obtained from orders@intechopen.com

Renewable and Sustainable Composites
Edited by António B. Pereira and Fábio A. O. Fernandes
p. cm.
Print ISBN 978-1-78984-215-9
Online ISBN 978-1-78984-216-6
eBook (PDF) ISBN 978-1-83962-860-3

For EU product safety concerns:
IN TECH d.o.o., Prolaz Marije Krucifikse Kozulić 3, 51000 Rijeka, Croatia, info@intechopen.com or visit our website at intechopen.com.

Meet the editors

António Pereira has a PhD in Mechanical Engineering and is currently the Director of the Centre for Mechanical Technology and Automation, Aveiro, Portugal. His main research area has been welding and the mechanics of composite materials. He has published 50 papers in SCI journals and has delivered 30 presentations at international conferences. His h-index at scopus is 17 with about 1000 citations.

Fábio Fernandes has a PhD in Mechanical Engineering and is a researcher at the University of Aveiro, Portugal. He has experience in material characterisation, Computer-aided Design/Engineering (CAD/CAE) and finite element analyses of real-world nonlinear problems. His research has resulted in several publications, including articles in high impact factor SCI journals. He has received many awards, most recently with the Mário Quartin Graça Scientific award, and has also received several honourable mentions.

Contents

Preface

The aim of this book is to present significant contributions in the field of composite structures based on sustainable solutions, including renewable materials. Although the content is based on several decades of research, the book also includes recent advances in this field.

The book is composed of seven chapters, starting with an introductory chapter, providing a broader introduction, supplementing the reader with further explanation of the purpose of the book, the nature of its content as well as the structure and its logic. The introductory chapter is followed by the presentation of methods to prepare and analyse composites, namely a review of the determinant factors that dictate the successful preparation of fibre-reinforced composites based on polymeric matrices and natural raw materials. This work is followed by a chapter on elastomeric composites filled with minerals at micro- and nanoscales. The next chapter attempts to answer another important challenge that society is facing: the production of clean energy. This chapter focuses on the development of 1D ZnO nanostructures to be employed in solar cells.

The next chapter presents a novel biocomposite developed in the context of one of the most important social challenges: the removal of pathogen colonies from contaminated waters. The work presented in this chapter do not only focus on the substitution of petroleum-derived materials by sustainable options. There is always an additional goal to answer other challenges and/or to improve the composite performance in a specific application.

The last three chapters are more focused on renewable and natural materials. First, the development of composites based on renewable, recyclable and natural materials, as well as biodegradable additives is presented. Finally, the book concludes with a chapter on the improvement of the mechanical properties of composites based on natural fibres for structural and biomedical applications.

António B. Pereira and Fábio A. O. Fernandes
TEMA – Centre of Mechanical Technology and Automation,
Department of Mechanical Engineering,
University of Aveiro,
Aveiro, Portugal

Section 1

The Importance of Composites in the World

Chapter 1

Introductory Chapter: The Importance of Composites in the World

António B. Pereira and Fábio A.O. Fernandes

The title of this book is composed by three well-known keywords. The first, **composites**, refer to something that is made up of several parts or elements, an overarching thematic. In fact, almost everything in the world is a composite, from the grain of sand to buildings or even the living beings. **Renewable** is a word that contains a fantastic concept, which we all aspire to: Anything renewable can be replaced or has an infinite source. When renewal is made in a timely manner, we can say that it is sustainable. This book provides ways and solutions for the creation and/or use of composites produced by **sustainable materials**.

Composites are currently used in a wide range of activity sectors, ranging from consumer electronics to the aeronautics and space industries. The excellent mechanical properties and low specific weight form the basis of the increasing volume of structural applications. An important milestone in the large-scale use of composite materials was the Airbus A380, the largest commercial aircraft in the world. Despite the progress achieved, there are significant obstacles to the generalization of structural applications of composites, not only in terms of high material costs but also due to the complexity of the mechanical behaviour. In fact, there is currently no sufficiently deep knowledge of the mechanisms of damage and ruin, especially with regard to composites made from renewable and sustainable materials.

Natural fibers are increasingly associated with attempts to replace other fibres and also with improvements in reinforcing structures [1]. For example, the incorporation of coconut fibres into a composite material of polyethylene and cork dust has already been studied, and an increase in tensile strength compared to unreinforced material has been found [2]. In another study, different laminates of fibreglass, hemp, flax and basalt were compared in flexural tests. The best performance was achieved by hybrid laminates in comparison with those made of only one type of fibers [3].

Today there are many studies in the field of composite materials with the goal to make the world greener. For instance, in the building sector, Dweib et al. [4] developed an all-natural composite structure for housing roof application. The materials used were resin based on soybean oil and natural fibers such as flax, cellulose, recycled paper and chicken feathers. These composite structures were compared with the wood structures currently used in construction and were completely able to effectively meet the requirements. These examples demonstrate the capability of sustainable composite materials as well as the breadth of exploitable applications.

Research teams across the globe are focusing on biocomposites, e.g. fiber-reinforced composites with both reinforcement and matrix from natural and renewable resources and also on hybrid solutions, consisting of natural fibers and oil-derived matrix or vice versa. Currently, hybrid solutions are more successful than fully green ones, usually due to cost, durability, performance and/or restrictions specific to the application. Nevertheless, fully green composites are gaining momentum in several industries, e.g., sustainable biocarbon fillers derived from wastes (biomass, industrial, food, etc.) have been showing a tremendous potential for the manufacturing of lightweight composite parts for the automotive industry [5].

From the perspective of mechanical engineering, the main values of a composite are its high strength coupled with its low specific weight, which allow it to be used in mobility applications, such as those related to transport. However, with the advent of nanotechnologies, enhancement additives such as nanomaterials and biodegradable ones, and technologies for composite manufacturing, have enabled new applications in the opposite spectrum of mechanical strength, for example, with new filtration products.

Additionally, in the context of mechanical engineering, another challenge is the smart integration of processing and manufacturing technologies to successfully join renewable materials into green biocomposite structures that can compete with current solutions in both performance and cost. Currently, compatibility of these materials in composite structures is a major challenge, with many researchers focusing on improving the interface/bonding between these materials and, thus, the mechanical performance of these structures.

Water scarcity and possible treatment and/or purification processes for the elimination of heavy metals or pathogens are current issues. In fact, climate change aggravates this reality. Today there are biocompatible and non-toxic composites formed of natural matrices to which nanoparticles are added, forming a biodegradable composite film which is expected to be low cost and an effective way to remove pathogens from water.

Circular economy is a hot ideal on the agenda of many organizations and governments, which is worth to strive for zero waste and total recyclability. The use of highly efficient composites with renewable, recyclable or degradable reinforcements contributes to reduce pollution, reducing waste and promoting green economy. Reusability is clearly a mandatory solution for sustainability. The use of nanoscale additives can make a huge contribution to the customization (specialization) of the composite and, for example, to its degradation for reuse of constituents.

Reutilization of wastes currently disposed by several industries must be assessed in order to determine the possibility of creating new by-products, which will have a positive impact in the entire value chain. An excellent example is agglomerated cork, a by-product from raw cork usually used only for wine stoppers and/or premium products. Additionally, the growth in the utilization and impact of composites from renewable and sustainable sources do not solely depend on industry stakeholders, and governmental policies play an important role on the fostering of eco-friendly composites, which should positively impact both society and economy.

In a society facing dramatic problems related to pollution and excessive consumption of fossil fuel-derived products with low levels of recyclability, the authors tried with this book to collect recent contributions that aim to answer some of these major issues.

Acknowledgements

This chapter is supported by the following projects: UID/EMS/00481/2019-FCT - FCT - Fundação para a Ciencia e a Tecnologia and CENTRO-01-0145-FEDER-022083 - Centro Portugal Regional Operational Programme (Centro2020), under the PORTUGAL 2020 Partnership Agreement, through the European Regional Development Fund. Fábio Fernandes fully acknowledges Fundação para a Ciencia e a Tecnologia (FCT) for the following research grant: CEECIND/01192/2017.

Author details

António B. Pereira* and Fábio A.O. Fernandes
TEMA – Centre of Mechanical Technology and Automation, Department of Mechanical Engineering, University of Aveiro, Aveiro, Portugal

*Address all correspondence to: abastos@ua.pt

References

[1] Fernandes FAO, Tavares JP, Alves de Sousa RJ, Pereira AB, Esteves JL. Manufacturing and testing composites based on natural materials. Procedia Manufacturing. 2017;**13**:227-234

[2] Fernandes EM, Correlo VM, Mano JF, Reis RL. Novel cork–polymer composites reinforced with short natural coconut fibres: Effect of fibre loading and coupling agent addition. Composites Science and Technology. 2013;**78**:56-62

[3] Petrucci R, Santulli C, Puglia D, Sarasini F, Torre L, Kenny JM. Mechanical characterization of hybrid composite laminates based on basalt fibres in combination with flax, hemp and glass fibres manufactured by vacuum infusion. Materials & Design. 2013;**49**:728-735

[4] Dweib MA, Hu B, O'Donnell A, Shenton HW, Wool RP. All natural composite sandwich beams for structural applications. Composite Structures. 2004;**63**(2):147-157

[5] Mohanty AK, Vivekanandhan S, Pin JM, Misra M. Composites from renewable and sustainable resources: Challenges and innovations. Science. 2018;**362**(6414):536-542

Section 2

Composite Preparation and Analysis

Chapter 2

Fiber-Matrix Relationship for Composites Preparation

Owonubi J. Shesan, Agwuncha C. Stephen, Anusionwu G. Chioma, Revaprasadu Neerish and Sadiku E. Rotimi

Abstract

Fiber-matrix interaction at the interphase is one very important property that is of great concern to all polymer scientists involved in polymer composites. Many of the failures can be traced to the type of interfacial interaction existing in the composites. That is why highlighting the factors that dictate the type of and the extent of interactions at the interphases become very necessary. Natural fiber polymer composites have found application in many fields of human endeavors. To continue this growth being experienced, the factors that determine the formation of good interaction at the interphase most be understood, so that they can be manipulated for a better result.

Keywords: interfacial interaction, polymer composites, natural fibers, surface treatment, filler dispersity

1. Introduction

Composites have been defined as materials made by mixing more than two chemically and physically dissimilar components together, physically or chemically, to form one new material [1–3]. In a composite, there is the continuous component known as the matrix and there is the discrete or discontinuous component called the fillers. In the composite material, both the matrix and the fillers come together to act as one material. The filler is the load bearing component of the composites while matrix bind the fillers together, which is the reinforcing material [4, 5]. There are different types of composites. These include: ceramics matrix composites (CMC), polymer matrix composites (PMC) and metal matrix composites (MMC) [6–8]. PMC's are of great interest around the world today with notable advantages that include its light weight, high stiffness, high strength and the ease of fabrication [9–12]. Polymer composites have been reported to be in use for numerous years with a market share reported to have increased tremendously in the last decade [9]. This has been attributed to the introduction of environmentally friendly natural fibers from plant sources. The processing and application of polymer composites requires a good understanding of all the factors that governs the relationship between all components that makes up the composite [4, 13–15]. The structure-property's relationship of any polymer composite is of fundamental importance right from their design which includes material selection. Many research have tried to explain the relationship through various experimental designs [16–18]. One important factor that was generally accepted to have pronounced impact on the processing and application properties is the interfacial interaction between the matrix and the fibers. This interfacial interaction can vary from mere physical interfacial to real chemical interaction [3, 11, 15, 19–22]. To properly understand which interaction has taken place, a good understanding of the materials involved is necessary. Therefore, this chapter seeks to give an insight explanation to the different types of matrixes and fibers available for polymer composite preparation and their possible interactions.

2. Fiber-matrix interaction in polymer composite

Polymer composites is an heterogeneous components material and so their properties, will be governed by factors such as component properties, composition of the individual components, chemical and physical structures and interfacial interactions [17, 23, 24]. Although, all the factors are equally important, the first three can be controlled before processing while the interfacial interactions can only be determined after processing and the extent of interaction can only be predicted using the necessary characterization techniques. Particle/particle interactions in polymer composites induce aggregation while filler/matrix interactions lead to an interphase development which yields characteristics which are modifications of those of the individual component [25]. In order to achieve a good migration of stress from the matrix to the fibers, a strong interaction is necessary, such as coupling that creates covalent bonds between the polymer matrix and the fibers. Although, secondary forces like Van der Waal forces or hydrogen bond may occur amongst the components [26–28].

For polymer composites filled with natural fibers, there are conditions that are necessary for the natural fiber polymer composites (NFPCs) to be able to carry out the objective principle effectively. These include (i) the length of fibers used must be sufficiently long, (ii) the orientation of the fibers must be in sync with that of the load, and (iii) the interfacial adhesion must be sufficiently strong enough [29, 30]. Therefore, the type of interfacial adhesion determines the performances of the NFPCs (such as barrier properties, mechanical and thermal properties). Other types of interactions reported include electrostatic forces, inter diffusion and mechanical interlocking.

3. Polymer matrix

In polymer composite preparation, the polymer matrix serves as the binding material (binder). It helps to hold the fillers in position and also helps to transmit stress within the material [31–33]. Therefore, if the interfacial interaction is poor, the transmission of an externally applied stress will be poorly done, leading to failures in the material. To ensure that there is a good stress transfer in polymer composites, the right polymer matrix is selected. Polymers are known for their unique properties which differ from one polymer to the other, even within the same group. Based on this, polymers are categorized on the basis of their chemical behavior (i.e. thermoplastic or thermosetting) or on the basis of their source (i.e. synthetic or natural). Thermoplastic are polymers that once processed and are harden to shape, can be reprocessed again and again as desired. Examples are polyethylene (PE), polypropylene (PP), poly(caprolactone) (PCL), poly(lactic acid) (PLA). Thermosetting are polymers that once processed at an elevated temperature and are set into shape, can harden and cannot be reprocessed again. Examples are

Biodegradable polymer matrices	
Natural	**Synthetic**
Polysaccharides—starch, cellulose, chitin Proteins—collagen/gelatin, casein, albumin, fibrogen Polyesters—polyhydroxyalkanoates (PHAs) Other polymers—lignin, lipids, shellac, natural rubber	Poly(amides) Poly(anhydrides) Poly(amide-enamines) Poly(vinyl alcohol) Poly(vinyl acetate) Polyesters—poly(glycolic acid), poly(actic acid), poly(caprolactone), poly(orthoesters), poly(ethylene oxides), poly(phosphazines)

Table 1.
Some common polymer matrices used for composite preparation [34].

polyesters, epoxies. Also these polymers are either synthesized or are obtained from nature. **Table 1** shows a list of polymer matrices grouped into natural and synthetic. All these influence the choice of polymer matrix.

4. Fibers

Fibers are one form of fillers that can be used to prepare polymer composites. Researchers have fibers in natural or synthetic forms [35–39]. Fibers such as aramid, carbon and glass are known as synthetic fibers. Due to increased environmental awareness, synthetic fibers are now being replaced with natural fibers which are more environmental friendly [5, 40–42]. Natural fibers have emerged as a viable alternative to their synthetic counterparts in polymer reinforcement, owing to the

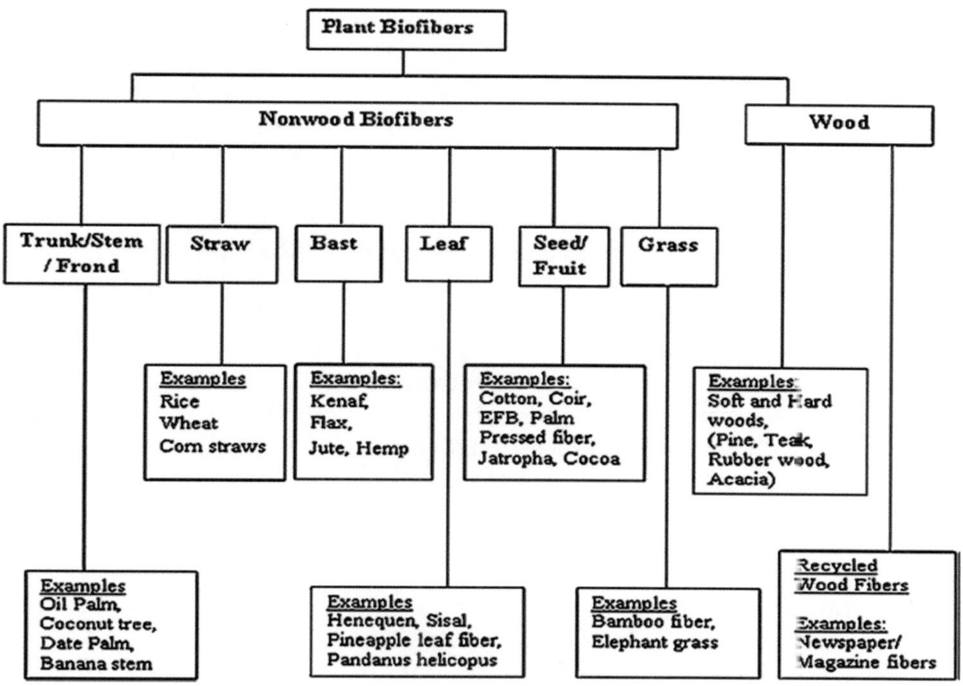

Figure 1.
Representation diagram showing the classification of natural fibers based on the part of plants they are extracted from [43].

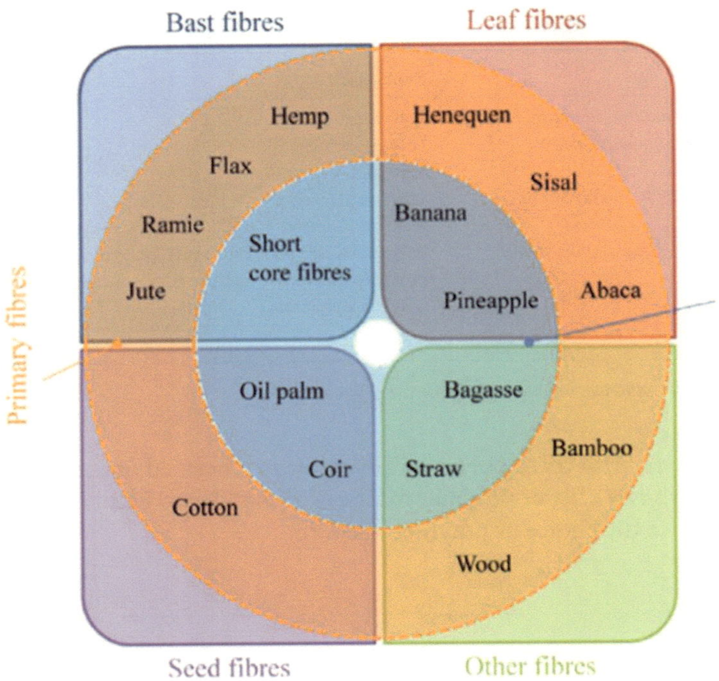

Figure 2.
Schematic classification of natural fibers as primary and secondary fibers [50].

large scale research output [40]. Natural fibers are obtained from natural source such as minerals, animals or plants [34]. Fibers from plant sources consist of cellulose, while those obtained from animal sources consist of proteins (wool, silk and hair). Natural fibers offer large diversity in terms of sources. Plant fibers are generally categorized based on their location with the plant as illustrated in **Figure 1** [4].

Fibers from plant source have been widely used in preparing polymer composites [44–49]. This is because of the availability and the easy of processing them. Plant fibers can further be classified a primary fibers, or secondary fibers [40, 50]. Primary fibers are the fibers obtained from plants, which were specifically grown for their fiber recourses. However, in cases where the fibers are by-products of other primary use of the plant, the fibers are referred to as secondary fibers (**Figure 2**).

The mechanical properties and morphology of these fibers are influenced greatly by the value of the ratio of fiber diameter to the cell wall thickness which by extension is influenced by the extraction methods used [40, 51, 52]. These factors dictate the type of interaction that will occur between the plant fibers and the polymer matrix, whether thermoplastic or thermoset [53, 54]. Natural fibers are good materials for many applications as they provide reinforcement properties at very low cost with low density, good strength and stiffness [8]. Their advantages over synthetic fibers have been well documented and they are based generally on environmental sustainability which include health and safety concerns [55, 56].

5. Preparation

With a careful selection of appropriate preparation, some level of desired interaction at the interface can be achieved. The processing of NFPCs has the ability to influence the final properties if carefully handled. Polymer composites can be prepared using the existing methods and technologies. These include injection molding, compression molding, blow molding, electro spinning, batch mixing or continuous mixing methods (extrusion), solution casting [34]. Different researchers have employed different methods to achieve their goal or specific property improvement. An experiment to toughen polylacticle/poly(butylenes succinate co-adipate) blends was reported [57]. This was achieved through melt blending (batch mixing) of the mixture in the presence of a reactive compatibilizer tripheny phosphate. Polypropylene/carbon nanotube composites were prepared using a twin screw extrusion mixer [58]. Their intention was to characterize the rheological behaviors and develop a model for the flow inside the machine. To investigate the micro mechanical characteristics of jute fiber/polypropylene composites, Liu et al. prepared the PP reinforced jute fiber Mat using film—stacking methods [7]. This was done by placing two layers of pre-dried non-woven jute mats in a mixture with three layers of PP films. In trying to prepare a novel composite films of polypyrrole-nanotube/polyaniline, Zhang et al., used facile solvent-evaporation method [59]. Also, recycling of biodegradable polymers for composites preparation has been done using extrusion processing [60]. The earlier mentioned methods of processing of polymer composites can be grouped into three primary systems, namely: solution blending, in-situ polymerization and melt blending [61].

5.1 Solution blending or casting

This involves the formation of the composites through dissolution of the polymer and fibers in a suitable solvent and subsequent drying of the solvent after the processing of the composites [62–64]. In solution casting, obtaining the right solvent that will dissolve the polymer and be removed with easy is a major task, although there are water-soluble polymers [65, 66]. Examples are polyvinyl pyrrolidone (PVP), polyvinyl alcohol (PVA), poly(ethylene oxide) (PEO) and poly(ethylene-co-vinyl acetate) (PEVA). Many others are dissolved using non-aqueous solvents such as chloroform, xylene, benzonitrile, tetrahydrofuran (THF), toluene, dimethylformanide (DMF) and acetone [67–69].

Solution casting helps to avoid thermo-mechanical degradation usually observed with the other methods of polymer processing which involve heating and vigorous mixing. It is used mostly for film formation. The amount of solvent used in this method makes solvent casting environmentally unsustainable. In addition, most of the organic solvents are hazardous to health and are temperature sensitive. Sur et al., in their work, prepared polysulfone nanocomposites using solution casting at elevated temperature [70]. To improve miscibility amongst the fibers and polymer matrix, the fibers were soaked using the solvent separately first before mixing with the dissolved polymer. In some cases, researchers coated the fibers with selected polymers to improve the fibers miscibility and interaction [71]. In general, solution casting is preferred for polymers with poor thermal stability and susceptible to thermo-mechanical degradation.

5.2 Melt blending

For the past decade now, melt blending has become a method of choice in the processing or preparation of natural fiber composites. It involves the heating of the polymer matrix up to 10–30°C above its melting point and then introduced the fiber into the molten polymer with thoroughly mixing under shear [72]. Melt blending has gained much ground in terms of acceptability, because of the existent compatibility with existing processing technology (such as injection molding and extrusion) [73–75]. These methods are environmentally friendly and do not involve the use of hazardous solvents. Many research works have reported the use of melt blending method [58, 76, 77]. One area of concern is the processing condition and the level of interaction between the fibers and the polymer matrix used. According to [61], to obtain good distribution of the fibers within the matrix, which is one of the conditions for improved mechanical and thermal properties, there should be a favorable enthalpy of interaction between the fibers and the polymer matrix. This results in good distribution of the fiber inside the matrix. In the absent of this favorable energy, the fibers will be poorly dispersed. The conditions for processing natural fiber composites will be discussed elaborately in subsequent sections.

5.3 In-situ polymerization

This technique involves the polymerization of a monomer in the presence of another polymer, mostly in small quantities [78]. In-situ polymerization has been described as one of the important methods for compatibilizing polymer blends [79]. It allows the formation of covalent bonding between constituents which can result into graft or block copolymers that ultimately results in the development of a stable interface [72]. In-situ polymerization is noted to yield specific properties with conventional melt blending methods. Furthermore, it allows the preparation of composites with high fiber weight fraction because the homogeneity of the resultant composites is much greater than that obtainable from melt blending and solution casting [80]. Most thermoset NFPCs are prepared using in-situ polymerization methods [17, 78, 81]. According to Bounor-Legaré et al. [82], subject to the nature and reactiveness of the organic or inorganic precursors and the processing factors, different types of functionality can be fashioned.

6. Factors to consider before and during the processing of natural fiber polymer composites

The preparation of natural fiber polymer composites with good strength is dependent largely on some factors mainly connected to (i) the fiber properties, (ii) the polymer matrix and (iii) the fiber-matrix interface properties. The strength and stiffness of any polymer composite is a direct function of the reinforcing fiber properties [83]. On the other hand, the matrix helps to keep in position the fibers and also helps in the transferal of load from fibers to fibers [84]. This segment dwells on the factors that contribute to the fiber-matrix relationship.

6.1 Fiber types and surface treatment

Natural fibers used for NFPC are abounded and can be sourced from different kind of plants and from any part of the plant. **Figure 2** shows the classification of NF as primary and secondary fibers.

Plant fibers contain primarily cellulose, hemicellulose and lignin [85, 86]. However, the component of interest is the cellulose. It is a linear polymer of D-glucose units that are linked by β-1,4-glycosidic bonds. They are hydrophilic with the hydroxyl groups in each unit available to form hydrogen bonds which could be inter or intra molecular. This property helps the cellulose chain to be more stiff and enhance its rigidity [87, 88]. Cellulose is a semi crystalline polymer. However, because cellulose is surrounded by cementitious materials such as lignin and hemicelluloses, the percentage content of cellulose in any plant fiber determines its usefulness [87]. **Table 2** gives a summary of % cellulose content in some selected plant fiber.

Fiber type	Cellulose (wt%)	Hemicellulose (wt%)	Lignin (wt%)
Abaca	12	56–70.2	5.6–12
Bagasse	17	32–55.2	19.9–25.3
Banana	9.4–22	60–65	5–32
Bamboo	11–17	26–43	21–31
Coir	4–6	32–46	32.7–45
Cotton	5.5–12.6	82–96	0.5–1
Flax	27.6–70	60–81	2–3
Hemp	6–70	68–92	2.9–13
Jute	26.5	45–84	5–26
Kapok	4	13.16–64	5.54–22
Kenaf	6.26–53	37–72	9–21
Pineapple	1.44	80–85	4.6–12.7
Ramie	24.5	68–91	0.6–9.25
Sisal	9.4–22	43–78	4–12.0

Table 2.
Percentage cellulose contents of some selected plants [18].

To increase this percentage of cellulose in the fibers, the material is subjected to different kinds of modification including alkaline treatment. Alkaline modification helps to remove the hemicelluloses, lignin and all other water soluble contents of the fibers, and by extension, increase the cellulose content [60, 89]. Some researchers have been able to extract nanocellulose crystals with improvement in the modification processes [90]. This includes the bleaching of the alkaline treated fibers and then subjecting them to acid hydrolysis, giving rise to better quality cellulose at the nanoscale [91–94]. The treatment given to the fibers confirm on them increased rigidity with cleaner surfaces which exposes more of the hydroxyl groups to any further chemical modification [95–97]. **Figure 3** shows SEM images of raw fibers and those treated at different condition. It can be seen that those treated with alkaline and then acid hydrolysis give a pulp like morphology.

These nanoscale cellulose fibers have been reported to lead to improved interfacial interaction [91–94, 96–98]. Although, fibers possess hydrophilic properties in nature and polymers are hydrophobic, to improve the interfacial interaction, further chemical modification of the fiber surface may be carried out. As earlier mentioned alkaline treated cellulose, especially the nanocellulose, have high concentration of hydroxyl groups on the surfaces that allow for their sites to chemically alter the natural fibers (NF). Different methods and strategies have been employed to achieve this by researchers as shown in the schematic diagram by [41, 73] given in **Figure 4**. Such chemical modification should be mild in order to prevent any deterioration of the other use properties.

Furthermore, the physical properties, morphology and even the nano structure of the cellulose polymers depend greatly on the origin of the natural plant fibers and the processes and procedures of extraction. The extraction of nanocellulose from sisal, pineapple leaf and coir has been carried out [73, 99]. The results showed rod-like structures that are more separated with long, flexible and entangled morphology, especially from the banana rachis. This confirmed that the morphology and other physical properties of cellulose crystals is a function of the source of fiber. Furthermore, Le Bras et al. concluded in their work that the degree of crystallinity of the nanocellulose and its crystal structure depends on the method

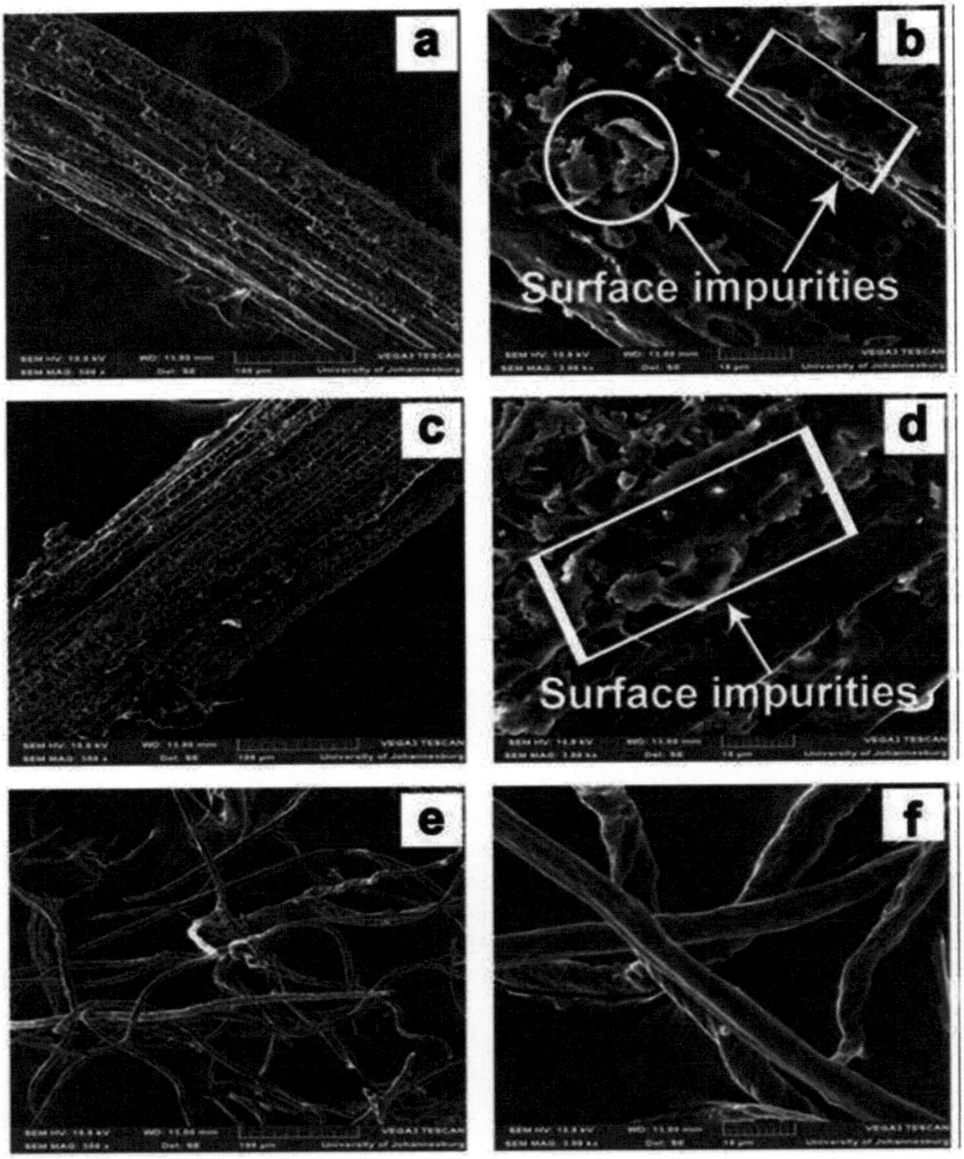

Figure 3.
SEM micrographs of untreated, alkaline treated and acid hydrolyzed treated fibers [73].

of extraction [100]. This was after the mechanical properties of the extracted nanocellulose were compared to each other and to those of their sources. Apart from this, the modified fibers have more uniform morphological structures. Also, there are reports that the thermal degradation property of natural fibers depends greatly on the level of modification given to the fibers (via acid hydrolysis, bleaching and alkaline treatment) and the source of the fibers [35]. This is because the nanocellulose obtained from such processes is more crystalline and rigid with uniform morphology. The factors that contribute to the variation in the lignocellulosic biomass include age of the plant, weather, plant type, soil nutrient, initial processes [87]. More recently, researchers have reported to have modified cellulose to allow for the introduction of functionalities to the polymer chains [41, 73]. The raw fiber can be pretreated using periodate oxidation and carboxymethylation techniques to introduce carboxyl and aldehyde functionalities which in turn could

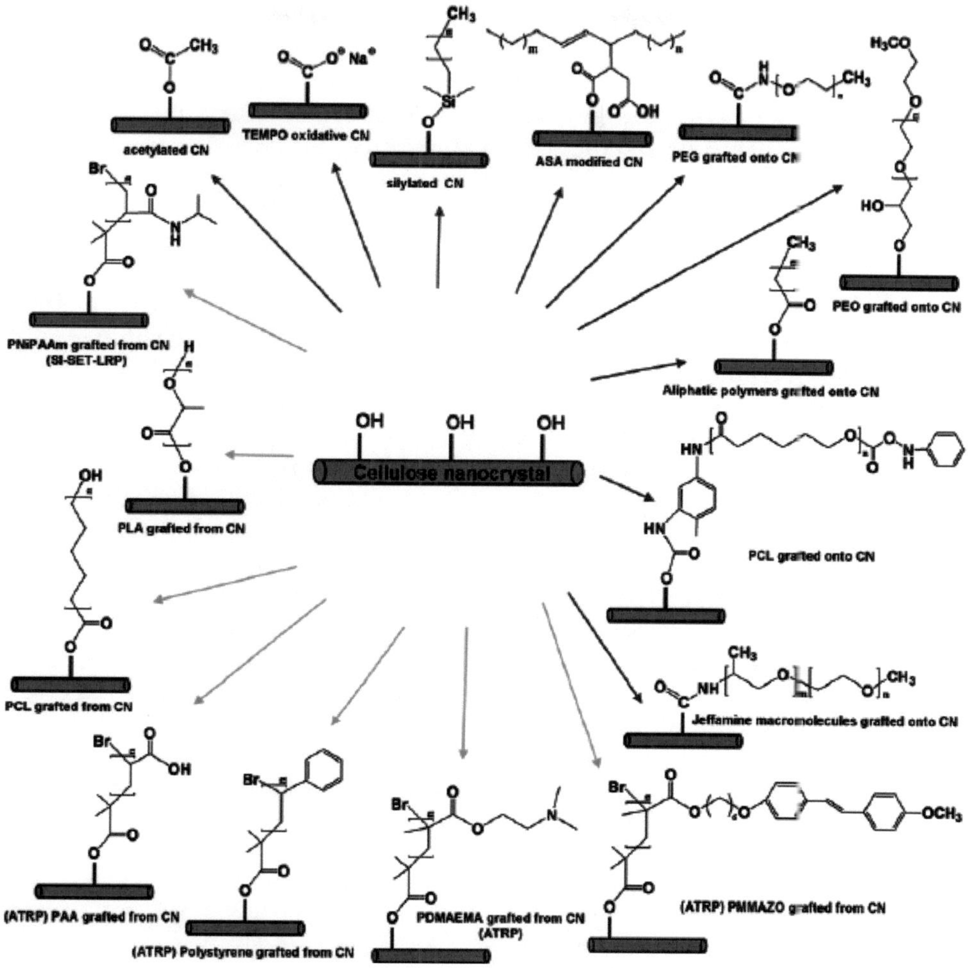

Figure 4.
Typical chemistry modification for cellulosic fibers [41, 73].

form covalent bonds with other functional groups such as amine. Fibers pretreated with bacterial method were modified with xyloglucan. This improved the wettability and cell adhesion for biomedical applications [101]. Chemical modification of cellulose fiber surfaces can lead to increased viscosity thereby reducing the shear thinning effect expected during processing. Therefore, fiber type, its modification and functionality can greatly improve fiber-matrix interfacial interaction, leading to improved performance properties.

NFPCs have exhibited significant potential for application in diverse sectors such as food packaging [102], fire retardant paperboard for semi structural applications [103], flexible printed electronics applications [104], etc. Nevertheless, there are numerous challenges to solve particularly associated with the development of proper large scale processing systems. The processing step of these materials is crucial because it is related to their final performance. Further studies are needed to prepare cellulose-based nanocomposites on industrial scale,

6.2 Dispersity of fibers

Natural fibers have the tendency to agglomerate in the polymer matrix as a result of formation of hydroxyl group's hydrogen bonding. This hydrogen bond formation leads to poor dispersion of the fibers within the matrix and by extension poor matrix-fiber interaction [73, 105]. The dispersity of the polar fiber which is hydrophilic in nature is worsened by the nonpolar hydrophobic nature of the polymer matrix. This factor has limited the extent of applicability of natural fiber polymer composites. In order to expand the areas of application, the fibers would need to be consistently and uniformly distributed within the matrix. To enhance the dispersity of these fibers in their polymer matrices, the modification and functionalization discussed earlier will become very important. The OH groups could be replaced by more hydrophobic organic moieties to help increase the affinity amongst the polymer matrix and the modified fibers. This will reduce the possibilities of hydrogen bonding and increase the dispersity in the matrix. The functionalization of the fibers may depend on the end use application. Yang et al., improved celluloses' suspension in aqueous media by acrylamide grafting on the surface by redox initiation grafting of extracted nanocellulose [106]. This led to improved self-assembly of the lyotropic state. Also, phosphorylation techniques with negatively charged phosphoric acid as phophoryl donor were used to modify the cellulose surfaces at the nanoscale level by [107]. Furthermore, with the right conditions, the hydroxyl groups were oxidized to dialdehyde using $NaIO_4$ [108]. Bae et al., used bimolecular nucleophilic substitution to replace the hydroxyl groups with long hydrophobia alkyl chains, in order to impart hydrophobicity to the cellulose polymer chain [109]. Also polycaprolactone diol (PCL) long chains were suggested for the surface modification of cellulose using what they referred to as click chemistry and esterification reactions [110]. Although it was reported that greater grafting yields were detected with the click chemistry when compared to the esterification method. The pretreatment given to the fibers also affects the extent to which the fiber surface can be modified.

6.3 Processing condition

During the processing of NFPCs, the processing temperature, mixing speed and residence time are all important for any melt blending process while for in-situ polymerization; the temperature and curing time are of importance [58, 78, 79]. For NFPCs to be prepared via melt blending, the temperature must not be too high to avoid the degradation of the natural fibers. Since natural fibers are thermally stable below 200°C, above this temperature, the integrity of the fibers cannot be guaranteed. Therefore, polymers with higher melting temperature may not be used in the preparation of NFPCs or alternative processing methods are applied. One factor to bear in mind is that the polymer must be molten enough to wet the fibers. This means that the Gibb's free energy of interaction tend towards negative. As the fibers are added, the viscosity increases. This increase is dependent on weight percentage of fibers added [13]. However, if the shear rates are increased it could minimize the effect of the increased viscosity [75, 77, 111–113]. Good fiber-matrix interaction can only be achieved if the polymer can wet the fiber surfaces properly. The resident time for processing the NFPC is also very important, one cannot use the same processing time as used for the neat polymer or blends, for NFPC, if other parameters are to remain the same. As mentioned earlier, the addition of fibers leads to change in viscosity; therefore NFPC will require more time to achieve good wettability. However, the option of increasing the shear rate or mixing speed might compensate for the change in viscosity. Therefore it may be possible to use the same resident time during processing. It must be mentioned that every option has its advantages and disadvantages. For example, longer time and increased mixing speed may lead to thermomechanical degradation [75].

6.4 Addition of other materials

In the preparation of composites, additives are added to help improve specific properties. For example, nanoparticles are added to enhance the mechanical and thermal properties through better crystallinity growth and stabilization of phase morphology [114, 115].

6.4.1 Compatibilizers

Compatibilizers are added to increase interfacial interactions amongst polymer-polymer in blends and polymer-fibers in composites [10, 41, 116]. The use of compatibilizers to improve interfacial interaction is widely reported [117]. In a series of works by Kamaker and other co-researchers [118–120], it was reported that Jute/PP composites' mechanical properties were improved drastically, when 3 wt% maleic anhydride grafted polypropylene (MAHgPP) was used to treat the Jute/PP composites as coupling agents. The tensile strength increased from 29.82 to 59.13 MPa, a 98% increase. Also, the bending strength was reported to increase from 49.97 to 87.66 MPa, a 75.4% increase. In related work, the tensile, flexural and dynamic strength of Jute/PP composites were enhanced by approximately 50% when the jute fibers are treated with 0.1 wt% MAHgPP solution of toluene, although the impact strength was negatively affected [121, 122]. Li et al., investigated two different kinds of silanes as coupling agents to treat sisal fibers surfaces [15]. The coupling agents were diluted in acetone to 6% concentration before use. A 24 h immersion of the sisal fibers was done using the coupling solution after which it was washed with acetone and dried up in the oven for 4 h at 50°C. Their results showed that the sisal surfaces were etched and they were very rough, which led the unbundling of the fiber bundle into smaller fibers, as seen in **Figure 3**. This increased the operative surface area presented for contact with the polymer matrix. Although, it was reported that the interaction at the interface was more of mechanical interlocking with less of chemical bonding. Therefore, the observed increase in load was due to frictional shear stress transferal across interface. Other coupling agents which had performed relatively well, have also been reported by other researchers [11, 123, 124]. However, to chemically modify the surface properties of the cellulose fibers to the extent that they can in turn influence the polymer composites properties significantly, it is suggested that the fibers should be modified to its nanoscale [125, 126].

6.4.2 Nanocellulose and nanoparticles

Nanocellulose particles are derived from rigorous chemical modification of plant fibers using a combination of alkaline and acidic solution in phases. At the nanoscale chemical modification is relatively easier; large numbers of the OH groups at the surface are more exposed; the amorphous components of the fibers have been eliminated, leaving only the crystalline part of the cellulose material. The nanoscale cellulose fibers when modified have wide application [127, 128]. Filler or additives can be described as materials which are added to the matrix in low percentages ranging from 0.1 to 5 wt% in order to improve performance and impact some special properties [129]. Some of the fillers are low-cost, allowing for a cost effective measure for the enhancement of performance properties. In addition, they can improve the processing of the material by controlling the viscosity of the resin [130]. Common fillers used in NFPCs include metal particles, nanosize cellulosic materials (CNCs), silica, nano-clay, maleic anhydride (MA) and carbon nano-tubes (CNT) [131]. Mohanty et al. showed that better interfacial adhesion existed amongst the fiber/matrix owing to filler addition in the NFPC as evident from the SEM micrographs of the fractured tensile samples; this resulted in the improvement of the mechanical properties [132]. Mechanical properties enhancement were resultant from the adding of fillers at low wt% which has led to the consistent assembly observed and enhanced stress transfer amongst the fiber and the matrix [1]. Meanwhile, at higher wt% of fillers, there was deterioration in the properties of the NFPCs, due to agglomeration of the fillers and the interfacial adhesion between the polymer matrix and fiber were noticed to be weak. Furthermore, it was suggested that the rise in tensile modulus and the fall in impact strength of the NFPCs with clay fillers were as a result of the polymer matrix improved cross-link density, which led to a decrease in the stiffness of the composite, hence reduced the impact strength [130, 133]. While in other studies, aluminum powder was used as fillers and in others, modified clay improved the impact strength of the composite, as the SEM micrographs revealed the existence of less voids and rise in density along with improvement in stiffness resultant from better interfacial adhesion amongst the fiber and the matrix in the presence of the modified clay [134, 135]. Other advantages of filler addition to sisal fiber reinforced polymer composites are increased moisture absorption or reduced water uptake as reported by [135, 136].

The addition of nanoparticles to NFPC to improve their interfacial properties has gained more popularity in recent time. Nanoparticles incorporation into polymer composites has been reported to improve their strength and young's modulus, as can be seen from earlier discussion. The addition of these nanoparticles at very low concentration of approximately 0.1–1.0% had influenced the mechanical properties of the NFPCs [137]. The modifications of the particles by some researcher have led to improved chemical interactions when added to the composites. Thereby improving interfacial bonding and by extension, mechanical and thermal properties [138–140]. Moreover, the addition of inorganic nanoparticles has helped to improve both thermal and mechanical [131]. This has influenced the crystallization process during processing of the composites. The improved crystallinity can encourage superior mechanical interlocking of the polymer-fiber interface.

6.4.3 Hybrid fibers

Hybrid fibers are a combination of two or more different fibers to make a composite. The addition of glass and/or carbon fibers to form hybrid fibers has also contributed to improving the composite properties. Allamraju et al., reported an increase in the compression and tensile strength of Jute/glass hybrid fibers epoxy composite, as a result of percentage increase in mass fraction of jute fibers [141]. Their results showed that the measured strengths increased as the fiber load increases from 6 to 9%, after which there was a decrease. The 9 wt% jute fibers that was reported as the optimal loading was attributed to higher stiffness of the composites and an improved fiber-matrix adhesion.

Hybrid composites are prepared by the combination of two or more different type, shape or size of reinforcement [142]. The crossbreed composite properties is completely dependent on many elements such as, extent of intermingling of fibers, fibers orientation, fiber surface roughness, compatibility between the fibers and their matrix, and the property of the individual fibers [143]. Recently, investigations on the properties of the crossbreed composites were centered on the natural/synthetic fibers, natural/natural fiber and natural/synthetic/additive modified reinforced polymer composites. Essentially for applications that required such hybrid/crossbreed composites with stiffness and high strength, but the employment of pure natural fiber polymer composites is difficult.

The popularity of these crossbreed types of composites is increasing rapidly owning to their capability to provide freedom to tailor the composites and achieving properties that cannot be acquired in composite encompassing only one kind of material [13, 144]. One of the major reasons for creating crossbreed NFPCs is to utilize the pluses of the fibers and lessen some drawbacks [145]. Another reason is based on economy, which is to decrease the cost of the finished composite product. Even though the usage of natural fibers in composites is less costly in comparison to the orthodox reinforcements, there are significant differences in the costs of natural fibers. Therefore, partially substituted cotton fibers in a composite made with wood fibers could offer very huge benefits in terms of the performance of the composite and also in the manufacture prices. The hybridization of wood flour- polypropylene composites with waste cone flour (20–40 wt%) was reported the composite was said to have been negatively affected in terms of their flexural properties and water resistance of the composite [146]. However, adding pine cone flour (10 wt %) to the composite revealed no substantial consequence on the measured properties, i.e. water absorption and flexural strength properties. This means it is economically advantageous. Further research showed that reducing the amount of pine cone flour added to the composite lead to positive improvement.

6.4.3.1 Natural/natural hybrids

Hybrid composites of sisal fibers and short banana fibers were also prepared using a polyester [147, 148]. From the result obtained, the tensile strength of the polyester composite was observed to rise as the volume fraction of the banana fiber was increased. On the other hand, the impact strength of the composites was affected negatively with increasing volume fraction of the banana fibers. Nevertheless, the impact strength improved with rise in total content of the fibers used. The observed properties were ascribed to two factors: (i) the lower microfibrillar angle of the banana fibers (11°) as compared to that of sisal (20°) and (ii) the better compatibility between the polymer matrix and banana fibers, which decreased the possibility of fiber pullout.

In the same vain, Venkateshwaran et al., prepared the same hybrid of banana and sisal fibers using epoxy resin in order to determine the optimal quantity of banana fibers with regard to its tensile properties [149]. The results revealed that about 50% of the complete fiber content added was good enough to impact significantly on the tensile strength of the composite. The differences in the morphologies of the different fibers have been shown to be of significant advantage when used as hybrid in a composite [142].

Furthermore, Fernandes et al., prepared hybrid composites of cork and coir fiber using high density polyethylene (HDPE) matrix [150]. With a coupling agent present, adding 10 wt% coir fibers to the composite caused a 30% rise in tensile maximum strength and a 39% rise in the tensile modulus. The overall effect of the coir fiber on the hybrid composites was very evident when compare to the single fiber composite, even with the coupling agent. Therefore, it can be said that the hybridization of NFPCs, most especially with natural/natural fibers, presented an efficient, sustainable and high economical way of improving the performance of the composites at a reduced cost. Also the addition of compatibilizers is still a necessity in order to create the needed covalent interactions required to enhance the performance characteristic property [151].

6.4.3.2 Natural/synthetic hybrid fibers

Natural fibers mixed with synthetic fibers hybridization could provide the desirable strength of a composite due to their synergistic effects [152]. Conversely, as a result of environmental impacts and disposal issues relating to the synthetic fibers in addition to hybrid strength requirements for specific applications, the final fiber ratio would be a 50:50 ratio (natural: synthetic). This ratio, offers a balance and intermediary mechanical properties in comparison to mechanical properties of either the synthetic or sisal fiber composites. Natural/natural fibers hybridization in a composite could be beneficial as a result of the possible disparity in cellulose quantity of the natural fibers used. This has a substantial effect on the composites mechanical properties and individual fibers have their distinctive characteristics which can be tailored for defined applications. A combination of such fibers together can bring about the needed properties for such application and still remain environmentally friendly. These biocomposites could be used in secondary/tertiary structures and other applications that require low stiffness and strength. Senthilkumar et al., highlighted that the enhancement in the mechanical properties of the hybrid composites was due to the strong impact from the type of polymer matrix used, fiber treatment technique, individual fiber loading and fiber choice [153]. Additionally, the enhancement in mechanical properties for natural/synthetic hybrid fiber reinforced composites was more significant than that of natural/natural hybrid fiber reinforced composites. It is the fiber/matrix interface adhesion that is responsible for the significant disparity observed in their mechanical properties for natural/synthetic fiber reinforced composites when compared to that of natural/natural hybrid fiber reinforced composite which displayed weak interfacial adhesion, more fiber pull out when under stress and uneven natural fiber distribution in the developed hybrid composites [153].

The general resolve for combining any two fiber kinds together in a single composite is to preserve the advantages of the two fibers in the new material and eliminate to the barest minimum their individual drawbacks [145, 154–157]. The synthetic fibers are known to possess good mechanical properties and thermal stability. They also lead to increase in overall weight of the composite, thereby eliminating the light weight advantage a complete natural fiber composite bears. However, the idea of combining natural fibers and synthetic fiber is influenced by the following parameters [152, 158, 159].

i. Relative amount of fiber

ii. Elastic properties of the fiber

iii. Failure stain ratio

iv. Fiber strength distribution

v. Degree of dispersion and uniformition

vi. Matrix properties

In summary, the addition of other component to NFPCs has led to reduction in some of the problems associated to their fiber-matrix interface. Compatibilization have helped reduce fiber agglomeration, water absorption and improved dispersity in the matrix.

7. Possible applications of NFPCs with improved fiber-matrix adhesion

The application of NFPCs comes with different requirements. These requirements are specific to application demands which include mechanical properties, thermal stability, transparence, conductivity and operational temperature. These factors are not unconnected to the factors considered before processing. For mechanical and thermal stability, the design of the NFPCs are considered from the point of improved interfacial interaction, using compartibilizers, fiber surface modification, addition of other chemical components and lastly reducing the fibers to nanosizes.

The automotive industry is one of the major end user playing a key role in the utilization of NFPCs. In the early '90s, Mercedes-Benz was the first, as a carmaker, to use NFPC, by building panels for doors with jute fibers [160]. This resulted in other car makers following suite, utilizing natural fibers comprising polymer composites for parcel shelves, headrests, upholstery, door panels, etc. The widespread adoption can be linked to the advantages of NFPCs, relating to the impact on the environmental, cost, elastic modulus and weight. No matter the applications, it was always a necessity to increase the mechanical properties of the composites via pre-treatment techniques (as discussed earlier). The fibers which have been treated were used in several ways, in order to obtain non-woven structures, mats, etc. [9]. Researchers have shown that subjecting natural fibers to some form of distinct treatment could lead to the development of high-quality composites with mechanical properties which are similar to those of glass fiber composites [160]. An outcome that would have ordinarily been very difficult to achieve because of natural fibers' hydrophilic nature which encourages water molecule absorption and agglomeration with no adhesion to the polymer matrix; This is the challenge and much works is ongoing to overcome this challenges [160, 161].

For conductivity, the addition of inorganic nanoparticles has been used and this has served a dual purpose of also improving the mechanical and thermal properties. Transparency is best impacted on NFPCs by the use of nanocellulose modified or not, with careful selection of the matrix. Some of the biodegradables are considered as having transparent properties including PLA. For water and gas permeability, applications are numerous with good water permeability. Such composites can be applied in water filtration processes, while for low water permeability composites; can be applied in packaging.

Researchers have reported positive impact of cellulose fibers on moisture and gas barrier performances for biodegradable polymers after modification with nanosize celluloses [162, 163]. Ambrosio-Martin and his colleagues prepared biodegradable composites of PLA using CNCs [164]. They reported improvements of barrier properties of oxygen and water which was due to the addition of well-dispersed OLLA-BCNCs. The researchers hypothesized that the nanofillers were able to make a tortuous path for the permeation of gas and water, thereby acting as blocking agents inside the polymeric matrix and hence, causing an increase to the barrier properties of the material. In these materials, the CNCs good level of dispersion within the polymer matrix, the morphology and its orientation led to enhanced tortuosity effects, thereby, heightening the barrier properties of the materials [165, 166]. Sanchez-Garcia and co-workers did similar work on PLA biodegradable materials and reported a decrease in the water permeability capacity and oxygen of approximately 82 and 90% respectively on addition of CNC to the PLA matrix [167]. In contrast, Espino-Pérez et al., highlighted results that were entirely opposite after compounding PLA with n-octadecyl-isocyanate-grafted-CNCs [168]. In their publication, it was clear that there was no reduction in oxygen permeability on adding CNCs into PLA. Apart from investigations about the tortuosity effects and barrier properties, researchers have tried to relate the improvement observed for the barrier properties to the materials crystallinity changes on addition of the nanofillers to the polymer matrix [169]. Fortunati et al, highlighted the improvements in barrier properties of PLA/CNC nanocomposites as well as increased crystallinity simultaneously, which was attributed to the addition of CNCs [170]. Espino-Pérez and co-workers investigated CNC/PLA nanocomposites using a high D-lactic acid content (a material which under normal processing conditions cannot crystallize) as matrix [171]. In conclusion, they indicated that the tortuosity effect of CNC on the oxygen barrier properties is limited. Although, after addition of modified CNS, significant improvement in the water vapor barrier properties was observed. They reported the swelling of unmodified CNCs due to absorption of water as a result of their hydrophilicity, which encouraged the pathway for mass transport and this property was not demonstrated when modified CNCs were employed because the surface modification caused the reduction in the hydrophilicity of the nanoparticles [172]. Follain et al., also reported that other elements can be considered to be significant in moisture and gas barrier performances of CNC-based nanocomposites [173]. They indicated that the formation of a 3D network and close interfacial adhesion between PCL chains and CNC can result in the matrix having structural defects, which encourage transfer of gas. Hence, their results highlighting the barrier properties of CNC-based nanocomposites allow for the conclusion that the tortuosity effect is influenced by CNCs, CNC surface chemistry, the structure of the nanocomposite. Also, that change(s) in crystallinity of the host matrix play a vital role on moisture and gas barrier performances of the material.

Biodegradation has been described as a vital prerequisite for biomedical materials, agricultural mulches and the packaging industry, as a result of the high level of consumption of these materials. Thus, the preparation of biodegradable polymers with improved properties is necessary but not without its own challenges. However, it will be exceptional to alleviate the concerns of landfills, chiefly in countries which are yet to adopt the technique of composting. A lot of research output has shown enhanced biodegradation for a number of polymer matrices when cellulosic fibers are added to them [174, 175]. However, the surface modification of the cellulose fibers negatively influences the matrix degradation because it reduces the number of OH group on the cellulose surface and so decreases the hydrophilicity [176]. Pinheiro and co-workers prepared poly (butylene adipate-co-terephthalate) PBAT-based composites with modified and unmodified cellulose fibers [174]. They presented from their results that the addition of unmodified cellulose fibers caused more weight reduction and this was attributed to the hydrophilic properties of the nanocrystals that hastened the hydrolysis of PBAT. In another related work, Monhanty et al, reported similar findings and also highlighted that the hydrophilic properties of the reinforcement encourages the degradation of the polymer [177]. The crystal size [178] and crystallinity of the polymer matrix [179, 180] have also been reported to playing a vital role in the degradation degree of the matrix, considering that regions which are crystalline are relatively unaffected by hydrolysis [178–182].

The use of other methods such as micro-fibrillation, laser, and ionomer to enhance the mechanical properties of materials have been reported. Choudhury and co-workers highlighted the capacity to improve the tensile and flexural properties of NFPCs by ionomer treatment of natural fibers [183]. The improvement was as a result of uniform stress dissemination and good dispersion of the fibers inside the matrix. This enhancement in mechanical properties by the microfibrillation was largely ascribed to the larger interaction observed between the polymer and fibers after the treatment by micro-fibrils and aggregates.

The selection of polymer matrix in all these areas of application becomes very important and sensitive. Although there are general purpose polymers with little or no health concerns for examples PP, PE, HDPE. When these polymers are compounded with NFs, their composites become more susceptible to microorganism attack, thereby making them biodegradable.

8. Conclusion

The dependence of performance properties on a strong fiber-matrix interface cannot be over emphasized. Fiber-matrix interfacial interactions are very important properties of all polymer composites. The performance properties are highly dependent on the kind of fiber-matrix interfacial bonds formed. These bonds can be physical or chemical in natural. The physical interaction includes interlocking between the matrix and the fibers, which is as a result of the rough edges of splits caused by the various surface treatments subjected to the fibers. While chemical interactions include the formation of bonds from the weak Van der Waal force to a strong covalent bond. The chemical bonds can be induced by (i) the type of surface treatment given to the fibers, which be either be a chemical or biological treatment and (ii) the deliberate addition of selected compatibilizers which confirm specific functionality on either the fibers or the polymer matrix. Furthermore, the addition of nanoparticles has been reported to help improve interfacial interactions through the direct participation in the crystallization processes which increases rigidity of the polymer matrix and thus enhances the physical interlocking at the interface. The use of hybrid fibers of polymer blends is another way of improving the interfacial interaction in NFPCs. Hybrid fibers can be natural/natural or natural/synthetic. While the natural/synthetic hybrid fibers clearly have better thermal and mechanical properties, their effect on the environment will always be of great concern. With appropriate surface treatments, natural/natural hybrid fibers have been reported to display improved thermal stability and good mechanical properties. Also, the use of polymer blends has resulted in improved interactions. The choice of processing methods and conditions can undermine the kind of and extent of interaction formed. While the in-situ polymerization is used for thermosets, it mostly leads to the formation of covalent interactions. The thermoplastics are prepared using melt blending which mostly gives rise to interlocking or at most Van der Waal force types of interaction at the interphases, although compatibilizer can be added the form covalent interactions. However, melt blending processes are more prone to thermomechanical degradation, which is likely to affect the blends performance properties. In conclusion, to ensure that the materials with the required properties are developed, it is necessary that the factors discussed above are properly considered.

Acknowledgements

The financial assistance of the University of Zululand and the National Research Foundation, South Africa through the South African Research Chair Initiative (SARChI) is hereby acknowledged. OSJ thanks the National Research Foundation (NRF) for a postdoctoral fellowship and funding under South African Research Chair for Nanotechnology.

Author details

Owonubi J. Shesan[1*], Agwuncha C. Stephen[2], Anusionwu G. Chioma[3], Revaprasadu Neerish[1] and Sadiku E. Rotimi[4]

1 Department of Chemistry, University of Zululand, Kwadlangezwa, Kwazulu-Natal, South Africa

2 Department of Chemistry, Ibrahim Badamasi Babangida University, Lapai, Nigeria

3 Department of Applied Chemistry, University of Johannesburg, Johannesburg, South Africa

4 Department of Chemical, Metallurgical and Materials Engineering, Institute of Nano Engineering Research (INER), Tshwane University of Technology, Pretoria, South Africa

*Address all correspondence to: oshesan@gmail.com

References

[1] Krishnan KA, Jose C, Rohith KR, George KE. Sisal nanofibril reinforced polypropylene/polystyrene blends: Morphology, mechanical, dynamic mechanical and water transmission studies. Industrial Crops and Products. 2015;**71**:173-184

[2] Kabir MM, Wang H, Lau KT, Cardona F. Chemical treatments on plant-based natural fibre reinforced polymer composites: An overview. Composites Part B: Engineering. 2012;**43**(7):2883-2892

[3] Merlini C, Soldi V, Barra GM. Influence of fiber surface treatment and length on physico-chemical properties of short random banana fiber-reinforced castor oil polyurethane composites. Polymer Testing. 2011;**30**(8):833-840

[4] Shekar HS, Ramachandra M. Green composites: A review. Materials Today: Proceedings. 2018;**5**(1):2518-2526

[5] Yuan X, Zhu B, Cai X, Qiao K, Zhao S, Yu J. Influence of different surface treatments on the interfacial adhesion of graphene oxide/carbon fiber/epoxy composites. Applied Surface Science. 2018;**458**:996-1005

[6] Jayaraman K. Manufacturing sisal-polypropylene composites with minimum fibre degradation. Composites Science and Technology. 2003;**63**(3-4):367-374

[7] Liu X, Dai G. Surface modification and micromechanical properties of jute fiber mat reinforced polypropylene composites. Express Polymer Letters. 2007;**1**(5):299-307

[8] Tao Y, Yan L, Jie R. Preparation and properties of short natural fiber reinforced poly (lactic acid) composites. Transactions of Nonferrous Metals Society of China. 2009;**19**:s651-s655

[9] La Mantia F, Morreale M. Green composites: A brief review. Composites Part A: Applied Science and Manufacturing. 2011;**42**(6):579-588

[10] Hamour N, Boukerrou A, Djidjelli H, Maigret J-E, Beaugrand J. Effects of MAPP compatibilization and acetylation treatment followed by hydrothermal aging on polypropylene alfa fiber composites. International Journal of Polymer Science. 2015;**2015**:9

[11] Keener TJ, Stuart RK, Brown TK. Maleated coupling agents for natural fibre composites. Composites Part A: Applied Science and Manufacturing. 2004;**35**(3): 357-362

[12] Kaur J, Lee JH, Shofner ML. Influence of polymer matrix crystallinity on nanocomposite morphology and properties. Polymer. 2011;**52**(19):4337-4344

[13] Väisänen T, Das O, Tomppo L. A review on new bio-based constituents for natural fiber-polymer composites. Journal of Cleaner Production. 2017;**149**:582-596

[14] Gao S-L, Mäder E. Jute/polypropylene composites I. Effect of matrix modification. Composites Science and Technology. 2006;**66**(7-8):952-963

[15] Li Y, Hu C, Yu Y. Interfacial studies of sisal fiber reinforced high density polyethylene (HDPE) composites. Composites Part A: Applied Science and Manufacturing. 2008;**39**(4):570-578

[16] Thakur VK, Singha AS, Kaur I, Nagarajarao RP, Liping Y. Studies on analysis and characterization of phenolic composites fabricated from lignocellulosic fibres. Polymers & Polymer Composites. 2011;**19**(6):505

[17] Vieille B, Taleb L. About the influence of temperature and matrix ductility on the behavior of carbon woven-ply PPS or epoxy laminates: Notched and unnotched laminates. Composites Science and Technology. 2011;**71**(7):998-1007

[18] Dunne R, Desai D, Sadiku R, Jayaramudu J. A review of natural fibres, their sustainability and automotive applications. Journal of Reinforced Plastics and Composites. 2016;**35**(13):1041-1050

[19] Nascimento DM, Almeida JS, Dias AF, Figueirêdo MCB, Morais JPS, Feitosa JP, et al. A novel green approach for the preparation of cellulose nanowhiskers from white coir. Carbohydrate Polymers. 2014;**110**:456-463

[20] Jain D, Vats S, Kumar Bera T. Micromechanical interactions and their relation to stress field for different classes of reinforced polymer composites. Materials Today: Proceedings. 2018;**5** (9, Part 3):19944-19953

[21] Kunanopparat T, Menut P, Morel M-H, Guilbert S. Plasticized wheat gluten reinforcement with natural fibers: Effect of thermal treatment on the fiber/matrix adhesion. Composites Part A: Applied Science and Manufacturing. 2008;**39**(12):1787-1792

[22] Alam A, Shubhra QT, Al-Imran G, Barai S, Islam M, Rahman MM. Preparation and characterization of natural silk fiber-reinforced polypropylene and synthetic E-glass fiber-reinforced polypropylene composites: A comparative study. Journal of Composite Materials. 2011;**45**(22):2301-2308

[23] Hancox N. Overview of effects of temperature and environment on performance of polymer matrix composite properties. Plastics Rubber and Composites Processing and Applications. 1998;**27**(3):97-106

[24] Kawai M, Yajima S, Hachinohe A, Kawase Y. High-temperature off-axis fatigue behaviour of unidirectional carbon-fibre-reinforced composites with different resin matrices. Composites Science and Technology. 2001;**61**(9):1285-1302

[25] Ferreira FV, Dufresne A, Pinheiro IF, Souza DHS, Gouveia RF, Mei LHI, et al. How do cellulose nanocrystals affect the overall properties of biodegradable polymer nanocomposites: A comprehensive review. European Polymer Journal. 2018;**108**:274-285

[26] Raju B, Hiremath S, Mahapatra DR. A review of micromechanics based models for effective elastic properties of reinforced polymer matrix composites. Composite Structures. 2018;**204**:607-619

[27] Mohebby B, Fallah-Moghadam P, Ghotbifar A, Kazemi-Najafi S. Influence of maleic-anhydride-polypropylene (MAPP) on wettability of polypropylene/wood flour/glass fiber hybrid composites. Journal of Agricultural Science and Technology. 2011;**13**:877-884

[28] Ku H, Wang H, Pattarachaiyakoop N, Trada M. A review on the tensile properties of natural fiber reinforced polymer composites. Composites Part B: Engineering. 2011;**42**(4):856-873

[29] Razak NIA, Ibrahim NA, Zainuddin N, Rayung M, Saad WZ. The influence of chemical surface modification of kenaf fiber using hydrogen peroxide on the mechanical properties of biodegradable kenaf fiber/poly (lactic acid) composites. Molecules. 2014;**19**(3):2957-2968

[30] da Luz FS, Ramos FJHTV, Nascimento LFC, da Silva Figueiredo AB-H, Monteiro SN. Critical length and interfacial strength of PALF and coir fiber incorporated in epoxy resin matrix. Journal of Materials Research and Technology. 2018;**7**(4):528-534

[31] Saba N, Jawaid M, Asim M. Recent advances in nanoclay/natural fibers hybrid composites. In: Nanoclay Reinforced Polymer Composites. Singapore: Springer; 2016. pp. 1-28

[32] Park SY, Lee BI, Jung ST, Park HJ. Biopolymer composite films based on κ-carrageenan and chitosan. Materials Research Bulletin. 2001;**36**(3-4):511-519

[33] Besson F, Budtova T. Cellulose ester-polyolefine binary blend: Morphological, rheological and mechanical properties. European Polymer Journal. 2012;**48**(5):981-989

[34] John MJ, Thomas S. Biofibres and biocomposites. Carbohydrate Polymers. 2008;**71**(3):343-364

[35] Abraham E, Deepa B, Pothan L, Jacob M, Thomas S, Cvelbar U, et al. Extraction of nanocellulose fibrils from lignocellulosic fibres: A novel approach. Carbohydrate Polymers. 2011;**86**(4):1468-1475

[36] Lee H-R, Kim K, Mun SC, Chang YK, Choi SQ. A new method to produce cellulose nanofibrils from microalgae and the measurement of their mechanical strength. Carbohydrate Polymers. 2018;**180**:276-285

[37] Seabra AB, Bernardes JS, Fávaro WJ, Paula AJ, Durán N. Cellulose nanocrystals as carriers in medicine and their toxicities: A review. Carbohydrate Polymers. 2017;**181**:514-527

[38] Boronat T, Fombuena V, Garcia-Sanoguera D, Sanchez-Nacher L, Balart R. Development of a biocomposite based on green polyethylene biopolymer and eggshell. Materials & Design. 2015;**68**:177-185

[39] Robles E, Fernandez-Rodriguez J, Barbosa AM, Gordobil O, Carreno NL, Labidi J. Production of cellulose nanoparticles from blue agave waste treated with environmentally friendly processes. Carbohydrate Polymers. 2018;**183**:294-302

[40] Bourmaud A, Beaugrand J, Shah DU, Placet V, Baley C. Towards the design of high-performance plant fibre composites: How can we best define the diversity and specificities of plant cell walls? Progress in Materials Science. 2018;**97**:347-408

[41] Yatigala NS, Bajwa DS, Bajwa SG. Compatibilization improves physico-mechanical properties of biodegradable biobased polymer composites. Composites Part A: Applied Science and Manufacturing. 2018;**107**:315-325

[42] Lila MK, Singhal A, Banwait SS, Singh I. A recyclability study of bagasse fiber reinforced polypropylene composites. Polymer Degradation and Stability. 2018;**152**:272-279

[43] Jawaid M, Abdul Khalil HPS. Cellulosic/synthetic fibre reinforced polymer hybrid composites: A review. Carbohydrate Polymers. 2011;**86**(1):1-18

[44] Hammiche D, Boukerrou A, Djidjelli H, Corre Y-M, Grohens Y, Pillin I. Hydrothermal ageing of alfa fiber reinforced polyvinylchloride composites. Construction and Building Materials. 2013;**47**:293-300

[45] Demir H, Atikler U, Balköse D, Tıhmınlıoğlu F. The effect of fiber surface treatments on the tensile and water sorption properties of polypropylene-luffa fiber composites. Composites Part A: Applied Science and Manufacturing. 2006;**37**(3):447-456

[46] Islam MS, Pickering KL, Foreman NJ. Influence of alkali treatment on the interfacial and physico-mechanical properties of industrial hemp fibre reinforced polylactic acid composites.

Composites Part A: Applied Science and Manufacturing. 2010;**41**(5):596-603

[47] Beaugrand J, Guessasma S. Scenarios of crack propagation in bast fibers: Combining experimental and finite element approaches. Composite Structures. 2015;**133**:667-678

[48] Ochi S. Mechanical properties of kenaf fibers and kenaf/PLA composites. Mechanics of Materials. 2008;**40**(4-5):446-452

[49] Huber T, Müssig J, Curnow O, Pang S, Bickerton S, Staiger MP. A critical review of all-cellulose composites. Journal of Materials Science. 2012;**47**(3):1171-1186

[50] Shah DU. Natural fibre composites: Comprehensive Ashby-type materials selection charts. Materials & Design. 2014;**62**:21-31

[51] Bourmaud A, Morvan C, Bouali A, Placet V, Perre P, Baley C. Relationships between micro-fibrillar angle, mechanical properties and biochemical composition of flax fibers. Industrial Crops and Products. 2013;**44**:343-351

[52] Gorshkova TA, Wyatt SE, Salnikov VV, Gibeaut DM, Ibragimov MR, Lozovaya VV, et al. Cell-wall polysaccharides of developing flax plants. Plant Physiology. 1996;**110**(3):721-729

[53] Meijer W, Vertregt N, Rutgers B, Van de Waart M. The pectin content as a measure of the retting and rettability of flax. Industrial Crops and Products. 1995;**4**(4):273-284

[54] Martin N, Mouret N, Davies P, Baley C. Influence of the degree of retting of flax fibers on the tensile properties of single fibers and short fiber/polypropylene composites. Industrial Crops and Products. 2013;**49**:755-767

[55] Reddy KO, Uma Maheswari C, Muzenda E, Shukla M, Rajulu AV. Extraction and characterization of cellulose from pretreated ficus (peepal tree) leaf fibers. Journal of Natural Fibers. 2016;**13**(1):54-64

[56] Cherian BM, Leão AL, de Souza SF, Costa LMM, de Olyveira GM, Kottaisamy M, et al. Cellulose nanocomposites with nanofibres isolated from pineapple leaf fibers for medical applications. Carbohydrate Polymers. 2011;**86**(4):1790-1798

[57] Ojijo V, Sinha Ray S, Sadiku R. Toughening of biodegradable polylactide/poly (butylene succinate-co-adipate) blends via in situ reactive compatibilization. ACS Applied Materials & Interfaces. 2013;**5**(10):4266-4276

[58] Thiébaud F, Gelin J. Characterization of rheological behaviors of polypropylene/carbon nanotubes composites and modeling their flow in a twin-screw mixer. Composites Science and Technology. 2010;**70**(4):647-656

[59] Zhang W-D, Xiao H-M, Fu S-Y. Preparation and characterization of novel polypyrrole-nanotube/polyaniline free-standing composite films via facile solvent-evaporation method. Composites Science and Technology. 2012;**72**(15):1812-1817

[60] Del Nobile MA, Conte A, Buonocore GG, Incoronato A, Massaro A, Panza O. Active packaging by extrusion processing of recyclable and biodegradable polymers. Journal of Food Engineering. 2009;**93**(1):1-6

[61] Ray SS. Clay-Containing Polymer Nanocomposites: From Fundamentals to Real Applications. Elsevier, United Kingdom: Newnes; 2013

[62] Churochkina N, Starodoubtsev S, Khokhlov A. Swelling and collapse of the gel composites based on neutral

and slightly charged poly (acrylamide) gels containing Na-montmorillonite. Polymer Gels and Networks. 1998;**6**(3-4):205-215

[63] Francis CW. Adsorption of polyvinylpyrrolidone on reference clay minerals. Soil Science. 1973;**115**(1):40-54

[64] Aranda P, Ruiz-Hitzky E. Poly (ethylene oxide)-silicate intercalation materials. Chemistry of Materials. 1992;**4**(6):1395-1403

[65] Wang S, Chen L, Tong Y. Structure-property relationship in chitosan-based biopolymer/montmorillonite nanocomposites. Journal of Polymer Science Part A: Polymer Chemistry. 2006;**44**(1):686-696

[66] Wang X, Liu B, Ren J, Liu C, Wang X, Wu J, et al. Preparation and characterization of new quaternized carboxymethyl chitosan/rectorite nanocomposite. Composites Science and Technology. 2010;**70**(7):1161-1167

[67] Chen G, Hao G, Guo T, Song M, Zhang B. Crystallization kinetics of poly (3-hydroxybutyrate-co-3-hydroxyvalerate)/clay nanocomposites. Journal of Applied Polymer Science. 2004;**93**(2):655-661

[68] Chen G, Hao G, Guo T, Song M, Zhang B. Structure and mechanical properties of poly (3-hydroxybutyrate-co-3-hydroxyvalerate) (PHBV)/clay nanocomposites. Journal of Materials Science Letters. 2002;**21**(20):1587-1589

[69] D'Amico DA, Manfredi LB, Cyras VP. Relationship between thermal properties, morphology, and crystallinity of nanocomposites based on polyhydroxybutyrate. Journal of Applied Polymer Science. 2012;**123**(1):200-208

[70] Sur G, Sun H, Lyu S, Mark J. Synthesis, structure, mechanical properties, and thermal stability of some polysulfone/organoclay nanocomposites. Polymer. 2001;**42**(24):9783-9789

[71] Huang X, Netravali AN. Characterization of nano-clay reinforced phytagel-modified soy protein concentrate resin. Biomacromolecules. 2006;**7**(10):2783-2789

[72] Robeson LM. Polymer Blends: A Comprehensive Review, Germany: Hanser Gardner; 2007

[73] Mondal S. Preparation, properties and applications of nanocellulosic materials. Carbohydrate Polymers. 2017;**163**:301-316

[74] Kiruthika AV. A review on physico-mechanical properties of bast fibre reinforced polymer composites. Journal of Building Engineering. 2017;**9**:91-99

[75] Abdel-Wahab AA, Ataya S, Silberschmidt VV. Temperature-dependent mechanical behaviour of PMMA: Experimental analysis and modelling. Polymer Testing. 2017;**58**:86-95

[76] Vasile C, Darie RN, Cheaburu-Yilmaz CN, Pricope G-M, Bračič M, Pamfil D, et al. Low density polyethylene-chitosan composites. Composites Part B: Engineering. 2013;**55**:314-323

[77] Kim H-S, Lee B-H, Choi S-W, Kim S, Kim H-J. The effect of types of maleic anhydride-grafted polypropylene (MAPP) on the interfacial adhesion properties of bio-flour-filled polypropylene composites. Composites Part A: Applied Science and Manufacturing. 2007;**38**(6):1473-1482

[78] Megiatto JD Jr, Silva CG, Ramires EC, Frollini E. Thermoset matrix reinforced with sisal fibers: Effect of the cure cycle on the properties of the

biobased composite. Polymer Testing. 2009;**28**(8):793-800

[79] Thakur VK, Thakur MK. Processing and characterization of natural cellulose fibers/thermoset polymer composites. Carbohydrate Polymers. 2014;**109**:102-117

[80] Spitalsky Z, Tasis D, Papagelis K, Galiotis C. Carbon nanotube-polymer composites: Chemistry, processing, mechanical and electrical properties. Progress in Polymer Science. 2010;**35**(3):357-401

[81] Rong MZ, Zhang MQ, Liu Y, Yang GC, Zeng HM. The effect of fiber treatment on the mechanical properties of unidirectional sisal-reinforced epoxy composites. Composites Science and Technology. 2001;**61**(10):1437-1447

[82] Bounor-Legaré V, Cassagnau P. In situ synthesis of organic-inorganic hybrids or nanocomposites from sol-gel chemistry in molten polymers. Progress in Polymer Science. 2014;**39**(8):1473-1497

[83] Sen T, Reddy HJ. Various industrial applications of hemp, kinaf, flax and ramie natural fibres. International Journal of Innovation, Management and Technology. 2011;**2**(3):192

[84] Joshi SV, Drzal LT, Mohanty AK, Arora S. Are natural fiber composites environmentally superior to glass fiber reinforced composites? Composites Part A: Applied Science and Manufacturing. 2004;**35**(3):371-376

[85] Ilyas RA, Sapuan SM, Ishak MR. Isolation and characterization of nanocrystalline cellulose from sugar palm fibres (*Arenga Pinnata*). Carbohydrate Polymers. 2018;**181**:1038-1051

[86] Bali G, Khunsupat R, Akinosho H, Payyavula RS, Samuel R, Tuskan GA, et al. Characterization of cellulose structure of populus plants modified in candidate cellulose biosynthesis genes. Biomass and Bioenergy. 2016;**94**:146-154

[87] Khoathane MC, Sadiku ER, Agwuncha CS. Surface modification of natural fiber composites and their potential applications. In: Surface Modification of Biopolymers. Hoboken, NJ, USA: John Wiley & Sons, Inc.; 2015. pp. 370-400

[88] Agwuncha SC, Anusionwu CG, Owonubi SJ, Sadiku ER, Busuguma AU, Ibrahim ID. Extraction of cellulose nanofibers and their eco-friendly polymer composites. In: Inamuddin et al., editors. Sustainable Polymer Composites and Nanocomposites. Springer International Publishing; 2018. p. VII, 2277

[89] Iwatake A, Nogi M, Yano H. Cellulose nanofiber-reinforced polylactic acid. Composites Science and Technology. 2008;**68**(9):2103-2106

[90] Abitbol T, Rivkin A, Cao Y, Nevo Y, Abraham E, Ben-Shalom T, et al. Nanocellulose, a tiny fiber with huge applications. Current Opinion in Biotechnology. 2016;**39**:76-88

[91] Alila S, Besbes I, Vilar MR, Mutjé P, Boufi S. Non-woody plants as raw materials for production of microfibrillated cellulose (MFC): A comparative study. Industrial Crops and Products. 2013;**41**:250-259

[92] Nascimento SA, Rezende CA. Combined approaches to obtain cellulose nanocrystals, nanofibrils and fermentable sugars from elephant grass. Carbohydrate Polymers. 2018;**180**:38-45

[93] Zhang X, Xu Y, Zhang X, Wu H, Shen J, Chen R, et al. Progress on the layer-by-layer assembly of multilayered polymer composites: Strategy, structural control and applications. Progress in Polymer Science. 2018;**89**:76-107

[94] Li Y, Zhu H, Shen F, Wan J, Lacey S, Fang Z, et al. Nanocellulose as green dispersant for two-dimensional energy materials. Nano Energy. 2015;**13**:346-354

[95] Zhang P, Li G. Advances in healing-on-demand polymers and polymer composites. Progress in Polymer Science. 2016;**57**:32-63

[96] Wang M, Bi W, Huang X, Chen DDY. Ball mill assisted rapid mechanochemical extraction method for natural products from plants. Journal of Chromatography A. 2016;**1449**:8-16

[97] Liu L, Ju M, Li W, Jiang Y. Cellulose extraction from *Zoysia* japonica pretreated by alumina-doped MgO in AMIMCl. Carbohydrate Polymers. 2014;**113**:1-8

[98] Zhang K, Sun P, Liu H, Shang S, Song J, Wang D. Extraction and comparison of carboxylated cellulose nanocrystals from bleached sugarcane bagasse pulp using two different oxidation methods. Carbohydrate Polymers. 2016;**138**:237-243

[99] Deepa B, Abraham E, Cordeiro N, Mozetic M, Mathew AP, Oksman K, et al. Utilization of various lignocellulosic biomass for the production of nanocellulose: A comparative study. Cellulose. 2015;**22**(2):1075-1090

[100] Le Bras D, Strømme M, Mihranyan A. Characterization of dielectric properties of nanocellulose from wood and algae for electrical insulator applications. The Journal of Physical Chemistry B. 2015;**119**(18):5911-5917

[101] Bodin A, Ahrenstedt L, Fink H, Brumer H, Risberg B, Gatenholm P. Modification of nanocellulose with a xyloglucan-RGD conjugate enhances adhesion and proliferation of endothelial cells: Implications for tissue engineering. Biomacromolecules. 2007;**8**(12):3697-3704

[102] Fortunati E, Luzi F, Yang W, Kenny JM, Torre L, Puglia D. Chapter 4—Bio-based nanocomposites in food packaging. In: Cerqueira MÂPR et al., editors. Nanomaterials for Food Packaging. United Kingdom: Elsevier; 2018. pp. 71-110

[103] Castro DO, Karim Z, Medina L, Häggström J-O, Carosio F, Svedberg A, et al. The use of a pilot-scale continuous paper process for fire retardant cellulose-kaolinite nanocomposites. Composites Science and Technology. 2018;**162**:215-224

[104] Hoeng F, Denneulin A, Bras J. Use of nanocellulose in printed electronics: A review. Nanoscale. 2016;**8**(27):13131-13154

[105] Andriy K. Predictive multiscale modeling of nanocellulose based materials and systems. IOP Conference Series: Materials Science and Engineering. 2014;**64**(1):012040

[106] Yang J, Ye DY. Liquid crystal of nanocellulose whiskers' grafted with acrylamide. Chinese Chemical Letters. 2012;**23**(3):367-370

[107] Kokol V, Božič M, Vogrinčič R, Mathew AP. Characterisation and properties of homo- and heterogenously phosphorylated nanocellulose. Carbohydrate Polymers. 2015;**125**:301-313

[108] Lu T, Li Q, Chen W, Yu H. Composite aerogels based on dialdehyde nanocellulose and collagen for potential applications as wound dressing and tissue engineering scaffold. Composites Science and Technology. 2014;**94**:132-138

[109] Bae JH, Kim SH. Alkylation of mixed micro-and nanocellulose

to improve dispersion in polylactide. Polymer International. 2015;**64**(6):821-827

[110] Benkaddour A, Jradi K, Robert S, Daneault C. Study of the effect of grafting method on surface polarity of tempo-oxidized nanocellulose using polycaprolactone as the modifying compound: Esterification versus click-chemistry. Nanomaterials. 2013;**3**(4):638-654

[111] Hufenbach W, Gude M, Böhm R, Zscheyge M. The effect of temperature on mechanical properties and failure behaviour of hybrid yarn textile-reinforced thermoplastics. Materials & Design. 2011;**32**(8):4278-4288

[112] Anderson BJ. Thermal stability and lifetime estimates of a high temperature epoxy by Tg reduction. Polymer Degradation and Stability. 2013;**98**(11):2375-2382

[113] González-Sánchez C, Fonseca-Valero C, Ochoa-Mendoza A, Garriga-Meco A, Rodríguez-Hurtado E. Rheological behavior of original and recycled cellulose-polyolefin composite materials. Composites Part A: Applied Science and Manufacturing. 2011;**42**(9):1075-1083

[114] Ibrahim ID, Jamiru T, Sadiku ER, Kupolati WK, Agwuncha SC, Ekundayo G. Mechanical properties of sisal fibre-reinforced polymer composites: A review. Composite Interfaces. 2016;**23**(1):15-36

[115] Ibrahim ID, Jamiru T, Sadiku RE, Kupolati WK, Agwuncha SC. Dependency of the mechanical properties of sisal fiber reinforced recycled polypropylene composites on fiber surface treatment, fiber content and nanoclay. Journal of Polymers and the Environment. 2017;**25**(2):427-434

[116] Thakur A, Purohit R, Rana R, Bandhu D. Characterization and evaluation of mechanical behavior of epoxy-CNT-bamboo matrix hybrid composites. Materials Today: Proceedings. 2018;**5**(2):3971-3980

[117] Zhou X, Yu Y, Lin Q, Chen L. Effects of maleic anhydride-grafted polypropylene (MAPP) on the physico-mechanical properties and rheological behavior of bamboo powder-polypropylene foamed composites. BioResources. 2013;**8**(4):6263-6279

[118] Karmaker A, Hoffmann A, Hinrichsen G. Influence of water uptake on the mechanical properties of jute fiber-reinforced polypropylene. Journal of Applied Polymer Science. 1994;**54**(12):1803-1807

[119] Karmaker A, Shneider J. Mechanical performance of short jute fibre reinforced polypropylene. Journal of Materials Science Letters. 1996;**15**(3):201-202

[120] Karmaker A. Effect of water absorption on dimensional stability and impact energy of jute fibre reinforced polypropylene. Journal of Materials Science Letters. 1997;**16**(6):462-464

[121] Gassan J, Bledzki AK. The influence of fiber-surface treatment on the mechanical properties of jute-polypropylene composites. Composites Part A: Applied Science and Manufacturing. 1997;**28**(12):1001-1005

[122] Gassan J, Bledzki AK. Influence of fiber surface treatment on the creep behavior of jute fiber-reinforced polypropylene. Journal of Thermoplastic Composite Materials. 1999;**12**(5):388-398

[123] Heinen W, Rosenmöller CH, Wenzel CB, de Groot HJM, Lugtenburg J, van Duin M. 13C NMR study of the grafting of maleic anhydride onto polyethene, polypropene, and ethene–propene copolymers. Macromolecules. 1996;**29**(4):1151-1157

[124] Ranganathan S, Baker W, Russell K, Whitney R. Peroxide-initiated grafting of maleic anhydride onto linear and branched hydrocarbons. Journal of Polymer Science Part A: Polymer Chemistry. 1999;**37**(20):3817-3825

[125] Sobolčiak P, Tanvir A, Popelka A, Moffat J, Mahmoud KA, Krupa I. The preparation, properties and applications of electrospun co-polyamide 6, 12 membranes modified by cellulose nanocrystals. Materials & Design. 2017;**132**:314-323

[126] Abdullah M, Nazir M, Raza M, Wahjoedi B, Yussof A. Autoclave and ultra-sonication treatments of oil palm empty fruit bunch fibers for cellulose extraction and its polypropylene composite properties. Journal of Cleaner Production. 2016;**126**:686-697

[127] Ferreira FV, Cividanes LDS, Brito FS, de Menezes BRC, Franceschi W, Simonetti EAN, et al. Functionalization of graphene and applications. In: Ferreira FV et al., editors. Functionalizing Graphene and Carbon Nanotubes: A Review. Cham: Springer International Publishing; 2016. pp. 1-29

[128] Ferreira FV, Francisco W, de Menezes BRC, Cividanes LDS, dos Reis Coutinho A, Thim GP. Carbon nanotube functionalized with dodecylamine for the effective dispersion in solvents. Applied Surface Science. 2015;**357**:2154-2159

[129] Saba N, Paridah M, Jawaid M. Mechanical properties of kenaf fibre reinforced polymer composite: A review. Construction and Building Materials. 2015;**76**:87-96

[130] Mallick PK. Fiber-Reinforced Composites: Materials, Manufacturing, and Design. USA: CRC press; 2007

[131] Ibrahim ID, Jamiru T, Sadiku ER, Kupolati WK, Agwuncha SC. Impact of surface modification and nanoparticle on sisal fiber reinforced polypropylene nanocomposites. Journal of Nanotechnology. 2016;**2016**:9

[132] Mohanty S, Verma SK, Nayak SK, Tripathy SS. Influence of fiber treatment on the performance of sisal-polypropylene composites. Journal of Applied Polymer Science. 2004;**94**(3):1336-1345

[133] Reddy Paluvai N, Mohanty S, Nayak S. Mechanical and thermal properties of sisal fiber reinforced acrylated epoxidized castor oil toughened diglycidyl ether of bisphenol A epoxy nanocomposites. Journal of Reinforced Plastics and Composites. 2015;**34**(18):1476-1490

[134] Wu L, Lu S, Pan L, Luo Q, Yang J, Hou L, et al. Enhanced thermal and mechanical properties of polypropylene composites with hyperbranched polyester grafted sisal microcrystalline. Fibers and Polymers. 2016;**17**(12):2153-2161

[135] Haldar P, Modak N, Sutradhar G. Comparative evaluation of mechanical properties of sisal-epoxy composites with and without addition of aluminium powder. Materials Today: Proceedings. 2017;**4**(2, Part A): 3397-3406

[136] Mohan T, Kanny K. Water barrier properties of nanoclay filled sisal fibre reinforced epoxy composites. Composites Part A: Applied Science and Manufacturing. 2011;**42**(4):385-393

[137] Ptak M, Kaczyński P, Wilhelm J, Margarido J, Marques P, Pinto S, et al. Graphene-enriched agglomerated cork material and its behaviour under quasi-static and dynamic loading. Materials. 2019;**12**(1):151

[138] Sadeghian R, Gangireddy S, Minaie B, Hsiao K-T. Manufacturing carbon nanofibers toughened polyester/glass

fiber composites using vacuum assisted resin transfer molding for enhancing the mode-I delamination resistance. Composites Part A: Applied Science and Manufacturing. 2006;**37**(10):1787-1795

[139] Dorigato A, Morandi S, Pegoretti A. Effect of nanoclay addition on the fiber/matrix adhesion in epoxy/glass composites. Journal of Composite Materials. 2012;**46**(12):1439-1451

[140] Pedrazzoli D, Pegoretti A, Kalaitzidou K. Synergistic effect of exfoliated graphite nanoplatelets and short glass fiber on the mechanical and interfacial properties of epoxy composites. Composites Science and Technology. 2014;**98**:15-21

[141] Allamraju KV. Study of mechanical behaviour of hybrid jute nano fiber composite. Materials Today: Proceedings. 2018;**5**(9):20750-20753

[142] Ashori A, Sheshmani S. Hybrid composites made from recycled materials: Moisture absorption and thickness swelling behavior. Bioresource Technology. 2010;**101**(12):4717-4720

[143] Saba N, Jawaid M, Alothman OY, Paridah M. A review on dynamic mechanical properties of natural fibre reinforced polymer composites. Construction and Building Materials. 2016;**106**:149-159

[144] Luo J, Semenikhin N, Chang H, Moon RJ, Kumar S. Post-sulfonation of cellulose nanofibrils with a one-step reaction to improve dispersibility. Carbohydrate Polymers. 2018;**181**:247-255

[145] Swolfs Y, Gorbatikh L, Verpoest I. Fibre hybridisation in polymer composites: A review. Composites Part A: Applied Science and Manufacturing. 2014;**67**:181-200

[146] Ayrilmis N, Buyuksari U, Dundar T. Waste pine cones as a source of reinforcing fillers for thermoplastic composites. Journal of Applied Polymer Science. 2010;**117**(4):2324-2330

[147] Idicula M, Malhotra S, Joseph K, Thomas S. Dynamic mechanical analysis of randomly oriented intimately mixed short banana/sisal hybrid fibre reinforced polyester composites. Composites Science and Technology. 2005;**65**(7-8):1077-1087

[148] Idicula M, Neelakantan N, Oommen Z, Joseph K, Thomas S. A study of the mechanical properties of randomly oriented short banana and sisal hybrid fiber reinforced polyester composites. Journal of Applied Polymer Science. 2005;**96**(5):1699-1709

[149] Venkateshwaran N, Elayaperumal A, Sathiya G. Prediction of tensile properties of hybrid-natural fiber composites. Composites Part B: Engineering. 2012;**43**(2):793-796

[150] Fernandes EM, Correlo VM, Mano JF, Reis RL. Novel cork-polymer composites reinforced with short natural coconut fibres: Effect of fibre loading and coupling agent addition. Composites Science and Technology. 2013;**78**: 56-62. Available from: https://www.sciencedirect.com/science/article/pii/S0266353813000389?via%3Dihub

[151] Ganesh BN, Rekha B. A comparative study on tensile behavior of plant and animal fiber reinforced composites. 2015;**1**(1):2015

[152] Swolfs Y, Gorbatikh L, Verpoest I. Stress concentrations in hybrid unidirectional fibre-reinforced composites with random fibre packings. Composites Science and Technology. 2013;**85**:10-16

[153] Senthilkumar K, Saba N, Rajini N, Chandrasekar M, Jawaid M, Siengchin S, et al. Mechanical properties evaluation of sisal fibre reinforced polymer composites: A review. Construction and Building Materials. 2018;**174**:713-729

[154] Kretsis G. A review of the tensile, compressive, flexural and shear properties of hybrid fibre-reinforced plastics. Composites. 1987;**18**(1):13-23

[155] Sevkat E, Liaw B, Delale F, Raju BB. Effect of repeated impacts on the response of plain-woven hybrid composites. Composites Part B: Engineering. 2010;**41**(5):403-413

[156] Wu Z, Yang C, Tobe Y, Ye L, Harada T. Electrical and mechanical characterization of hybrid CFRP sheets. Journal of Composite Materials. 2006;**40**(3):227-244

[157] Czél G, Wisnom M. Demonstration of pseudo-ductility in high performance glass/epoxy composites by hybridisation with thin-ply carbon prepreg. Composites Part A: Applied Science and Manufacturing. 2013;**52**:23-30

[158] Fukuda H, Chou TW. Stress concentrations in a hybrid composite sheet. Journal of Applied Mechanics. 1983;**50**(4a):845-848

[159] Zweben C. Tensile strength of hybrid composites. Journal of Materials Science. 1977;**12**(7):1325-1337

[160] Marsh G. Next step for automotive materials. Materials Today. 2003;**6**(4):36-43

[161] Prachayawarakorn J, Khunsumled S, Thongpin C, Kositchaiyong A, Sombatsompop N. Effects of silane and MAPE coupling agents on the properties and interfacial adhesion of wood-filled PVC/LDPE blend. Journal of Applied Polymer Science. 2008;**108**(6):3523-3530

[162] Ferrer A, Pal L, Hubbe M. Nanocellulose in packaging: Advances in barrier layer technologies. Industrial Crops and Products. 2017;**95**:574-582

[163] Nair SS, Zhu J, Deng Y, Ragauskas AJ. High performance green barriers based on nanocellulose. Sustainable Chemical Processes. 2014;**2**(1):23

[164] Ambrosio-Martín J, Fabra M, Lopez-Rubio A, Lagaron J. Melt polycondensation to improve the dispersion of bacterial cellulose into polylactide via melt compounding: Enhanced barrier and mechanical properties. Cellulose. 2015;**22**(2):1201-1226

[165] Espino-Pérez E, Bras J, Almeida G, Relkin P, Belgacem N, Plessis C, et al. Cellulose nanocrystal surface functionalization for the controlled sorption of water and organic vapours. Cellulose. 2016;**23**(5):2955-2970

[166] Follain N, Belbekhouche S, Bras J, Siqueira G, Marais S, Dufresne A. Water transport properties of bio-nanocomposites reinforced by Luffa cylindrica cellulose nanocrystals. Journal of Membrane Science. 2013;**427**:218-229

[167] Sanchez-Garcia MD, Lagaron JM. On the use of plant cellulose nanowhiskers to enhance the barrier properties of polylactic acid. Cellulose. 2010;**17**(5):987-1004

[168] Espino-Pérez EE, Bras J, Ducruet V, Guinault A, Dufresne A, Domenek S. Influence of chemical surface modification of cellulose nanowhiskers on thermal, mechanical, and barrier properties of poly(lactide) based bionanocomposites. European Polymer Journal. 2013;**49**(10):3144-3154

[169] Charlon S, Follain N, Chappey C, Dargent E, Soulestin J, Sclavons M, et al. Improvement of barrier properties of bio-based polyester nanocomposite membranes by water-assisted extrusion. Journal of Membrane Science. 2015;**496**:185-198

[170] Fortunati E, Peltzer M, Armentano I, Jiménez A, Kenny JM. Combined effects of cellulose nanocrystals and

silver nanoparticles on the barrier and migration properties of PLA nano-biocomposites. Journal of Food Engineering. 2013;**118**(1):117-124

[171] Espino-Pérez E, Bras J, Almeida G, Plessis C, Belgacem N, Perré P, et al. Designed cellulose nanocrystal surface properties for improving barrier properties in polylactide nanocomposites. Carbohydrate Polymers. 2018;**183**:267-277

[172] Tang J, Sisler J, Grishkewich N, Tam KC. Functionalization of cellulose nanocrystals for advanced applications. Journal of Colloid and Interface Science. 2017;**494**:397-409

[173] Follain N, Belbekhouche S, Bras J, Siqueira G, Chappey C, Marais S, et al. Tunable gas barrier properties of filled-PCL film by forming percolating cellulose network. Colloids and Surfaces A: Physicochemical and Engineering Aspects. 2018;**545**:26-30

[174] Pinheiro I, Ferreira F, Souza D, Gouveia R, Lona L, Morales A, et al. Mechanical, rheological and degradation properties of PBAT nanocomposites reinforced by functionalized cellulose nanocrystals. European Polymer Journal. 2017;**97**:356-365

[175] Garcia-Garcia D, Lopez-Martinez J, Balart R, Strömberg E, Moriana R. Reinforcing capability of cellulose nanocrystals obtained from pine cones in a biodegradable poly(3-hydroxybutyrate)/poly(ε-caprolactone) (PHB/PCL) thermoplastic blend. European Polymer Journal. 2018;**104**:10-18

[176] Brand J, Pecastaings G, Sèbe G. A versatile method for the surface tailoring of cellulose nanocrystal building blocks by acylation with functional vinyl esters. Carbohydrate Polymers. 2017;**169**:189-197

[177] Mohanty S, Nayak S. Biodegradable nanocomposites of poly (butylene adipate-co-terephthalate) (PBAT) and organically modified layered silicates. Journal of Polymers and the Environment. 2012;**20**(1):195-207

[178] Kanesawa Y, Tanahashi N, Doi Y, Saito T. Enzymatic degradation of microbial poly(3-hydroxyalkanoates). Polymer Degradation and Stability. 1994;**45**(2):179-185

[179] Bahari K, Mitomo H, Enjoji T, Yoshii F, Makuuchi K. Degradability of poly (3-hydroxybutyrate) and its copolymer grafted with styrene by radiation. Polymer Degradation and Stability. 1998;**61**(2):245-252

[180] Tomasi G, Scandola M, Briese BH, Jendrossek D. Enzymatic degradation of bacterial poly (3-hydroxybutyrate) by a depolymerase from pseudomonas lemoignei. Macromolecules. 1996;**29**(2):507-513

[181] El-Hadi A, Schnabel R, Straube E, Müller G, Henning S. Correlation between degree of crystallinity, morphology, glass temperature, mechanical properties and biodegradation of poly (3-hydroxyalkanoate) PHAs and their blends. Polymer Testing. 2002;**21**(6):665-674

[182] Arrieta MP, Fortunati E, Dominici F, López J, Kenny JM. Bionanocomposite films based on plasticized PLA-PHB/cellulose nanocrystal blends. Carbohydrate Polymers. 2015;**121**(Supplement C):265-275

[183] Choudhury A. Isothermal crystallization and mechanical behavior of ionomer treated sisal/HDPE composites. Materials Science and Engineering A. 2008;**491**(1-2):492-500

Chapter 3

Scanning Probe Microscopy of Elastomers with Mineral Fillers

Hammat H. Valiev, Alexander N. Vlasov, Yury V. Kornev, Yuliya N. Karnet, Nikolay A. Semenov and Oleg B. Yumaschev

Abstract

The results of a comprehensive study of the newly synthesized elastomeric composites filled with micro- and nanoscale modified shungite and also taurit, diatomit, and neosyl fillers are presented. The surface structure study of the prepared composites was conducted using scanning probe microscopy. The use of microscopy allowed visualization of the distribution patterns of filler aggregates and agglomerates in composites. The morphology and micro-nanometer size ranges of these aggregates in the synthesized materials are determined. The proposed method of grinding shungite, taurit, diatomit, and neosyl fillers allows significantly increasing the strength characteristics of these composites. The correlation between the reinforcement of the elastic-strength properties and the distribution of the used fillers in the rubber matrix was established.

Keywords: composites, synthetic rubber, mineral fillers, scanning probe microscopy, physico-mechanical properties

1. Introduction

Considering the high demand for composites made of elastomers with active fillers [1–4], it is necessary to search for renewable and sustainable fillers that are not inferior to those of traditional in reinforcing effect. Of particular interest are mineral fillers, which are fairly common and affordable raw materials with low cost and environmental friendliness. We investigated the use of minerals: shungite, taurit, diatomit, and a product of rice husk processing—neosyl—of various degrees of dispersion as fillers for elastomeric composites based on industrial synthetic rubber SBR-30 ARK. The use of scanning probe microscopy (SPM) allows us to study the surface structure of these composites and visualize the features of the distribution of macro- and nanofillers in the rubber matrix. Simultaneous testing of the elastic-strength properties of such materials is also important for establishing a correlation with the SPM data and revealing the nature of the reinforcement of composites. To increase the strength characteristics of the materials obtained, additional grinding of the used mineral fillers was required. These complex studies made it possible to establish the possibility of using these minerals as active fillers of the studied composites.

2. Experimental procedure and materials

Shungite is a mineral substance consisting of silicate particles dispersed in amorphous carbon mixed with inorganic substances. The chemical composition of used shungites is presented in **Table 1**.

New model samples of composites based on butadiene-styrene rubber (SBR) were selected in the ratio of the components in weight percentage (wt %): SBR-30ARK, 60 wt. %; shungit filler, 39 wt.%; and vulcanizing group, 1 wt.%. Nanoshungit fillers were produced by milling the initial mineral powder in a medium of isopropyl alcohol on a planetary ball mill RM 100 (Retsch, Germany). The functionalization of nanoshungit by organic modifier organosilanes was made directly during the grinding in the mill. Organosilanes used in parts 1.5 wt. % were:

- "TESPT": bis(3-triethoxysilylpropyl) tetrasulfide (C18H42O6S4Si2) STRUKTOL® SCA98
- "Si-264": 3-thiocyanatopropyltriethoxysilane (C10H21NO3SSi) STRUKTOL® SCA 984
- "Glymo": gamma-glycidoxypropyltrimethoxysilane (C9H20O5Si) STRUKTOL® SCA 960
- "Thiol": 3-mercaptopropyltriethoxysilane (C9H22O3SSi) STRUKTOL® SCA 989

Industrial products of Koksu shungit rocks of Kazakhstan are marketed under the brand name "Taurit." Taurit is a mineral with both organic and silicate parts in its structure (**Table 2**). It contains globular carbon with a metastable supramolecular structure of the siliceous or carbonate type. It has been established that the use of taurit in the composition of rubbers allows not only to reduce their cost but also to obtain improvement in some indicators technological and technical properties of rubbers.

Diatomites are sedimentary rocks of marine or lake genesis; more than 50% consist of siliceous, opal shells of microscopic algae—diatoms. The chemical composition and physical properties of diatomites (Karelia, Russia) depend on the species composition of the rock-forming diatoms and sedimentation conditions, determined by geological, geomorphological, and climatic factors (**Table 3**). Diatomite bulk sorbent is a highly porous material obtained by grinding diatom rock.

Neosyl-120 is a product of rice husk high-temperature processing and low access of oxygen, with grinding to sizes of the order of hundreds of micrometers with a wide distribution of particle size. The chemical composition of Neosyl-120 is shown in **Table 4**.

The grinding of the initial microstructural powders of shungit, taurit, diatomit, and Neosyl-120 to nanoscale was performed using a PM100 planetary ball mill (Retsch, Germany). The dispersion of fillers and blending of all ingredients of elastomers were carried out in a laboratory mixer HAAKE Rheomix

Substance	SiO_2	TiO_2	Al_2O_3	FeO	MgO	CaO	Na_2O	K_2O	S	C	H_2O
%	57.0	0.2	4.0	2.5	1.2	0.3	0.2	1.5	1.2	29.0	4.2

Table 1.
The chemical composition of shungites of the Zazhoginsky deposit type 3 (LTD "Carbon-Shungit" Karelia, Russia).

Element	TS %	Element	TS (%)
C	4–6	Mn	0.1
SiO_2	65–70	Ba	0.06
Al_2O_3	6–8	Zr	0.05
CaO	5–6	Sr	0.04
Fe_2O_3	3–4	V	0.015
Ti	0.5	B	0.01
K_2O	0.5	Zn	60 ppm
Na_2O	0.5	Ni	50 ppm

Table 2.
The average chemical composition of Koksu shale taurit [5].

Substance	SiO_2	Al_2O_3	Fe_2O_3	MgO	$K_2O + Na_2O$	Impurities
%	75	10	6	1	2	6

Table 3.
Chemical composition of diatomite sorbent [6].

Substance	SiO_2	Al_2O_3	FeO	CaO	MgO	K_2O	Na_2O	C	P_2O_5	Cl
%	92	0.5	0.2	1.1	0.5	3.0	0.1	1.5	0.3	0.1

Table 4.
Chemical composition of Neosyl-120.

No	Ingredients	Ш-940	Ш-941	Ш-942	Ш-943	Ш-944	Ш-945	Ш-946	Ш-947	Ш-948
1	SBR-30ARK	100	100	100	100	100	100	100	100	100
2	Sulfur SulfenamideC Altax	4.7	4.7	4.7	4.7	4.7	4.7	4.7	4.7	4.7
3	Zinc oxide	3	3	3	3	3	3	3	3	3
4	Norman 346	7	7	7	7	7	7	7	7	7
5	Stearin technical	2	2	2	2	2	2	2	2	2
6	Taurit	0	65	0	0	0	0	0	0	0
7	Taurit (nano)	0	0	65	0	0	0	0	0	0
8	Sorbent diatomite (initial)	0	0	0	65	0	0	0	0	0
9	Sorbent diatomite (nano)	0	0	0	0	65	0	0	0	0
10	Diatomite (initial)	0	0	0	0	0	65	0	0	0

No	Ingredients	III-940	III-941	III-942	III-943	III-944	III-945	III-946	III-947	III-948
11	Diatomite (nano)	0	0	0	0	0	0	65	0	0
12	Neosyl 120	0	0	0	0	0	0	0	65	0
13	Neosyl 120 (nano)	0	0	0	0	0	0	0	0	65

Table 5.
The composition of the studied rubber mixtures III-940–III-948 (weight percentage).

(Germany). Kinetics of vulcanization was investigated by analyzer RPA 2000 (Alpha Technologies, England). The mixture optimum curing was determined from obtained graphs. The composition of the synthesized elastomeric materials III-940–III-948 with taurit, diatomit, and Neosyl-120 fillers is presented in **Table 5**.

The study of the obtained samples of composites was carried out on an easyScan scanning probe microscope (Nanosurf, Switzerland), which operated in contact or semicontact modes in air at room temperature [7]. The modulation of force or phase contrast was also used to obtain the material contrasts of the composites under study. The processing of the obtained SPM images was carried out using the computer program SPIP (Image Metrology, Denmark). Scanning electron microscopy (SEM) images of the original Neosyl-120 powders were obtained using the electron microscope Jeol JSM-6510LV. Studies of physico-mechanical properties of the composites with these micro- and nanofillers were conducted on a universal tensile testing machine UTS-10 (Zwick Roell, Germany) and tensile testing machine with a pendulum force meter RMI-60 (ZIM, Tochmashpribor, Russia). The five repetitions for each type in tensile tests were done, and the mean values in the figures are presented. Dimensions of the samples of double-sided composite blades (type 1) correspond to the drawing and the table of reference [8].

3. Experimental results

SPM images of the surface structure of the synthesized composites with shungit unmodified and modified by organosilanes [9] are shown in **Figures 1–4**.

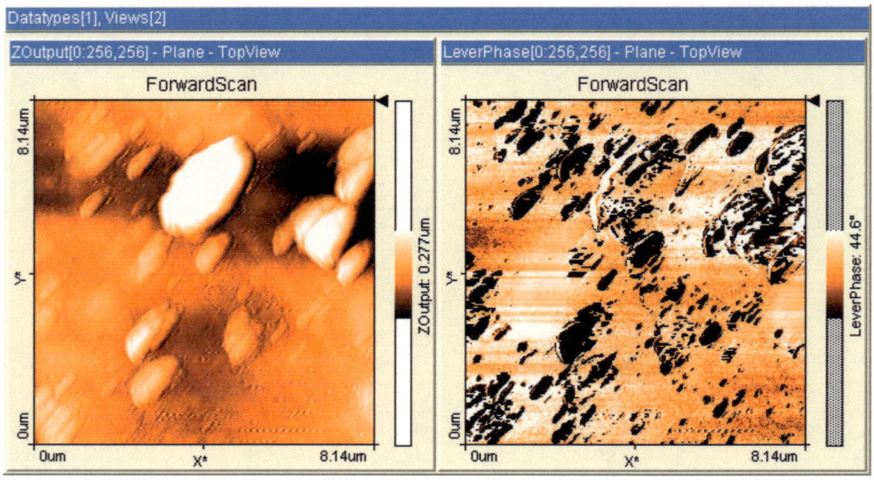

Figure 1.
SPM image of distribution in the rubber of original schungit. Scan 8.14 × 8.14 μm². Left—topography; and right—phase contrast.

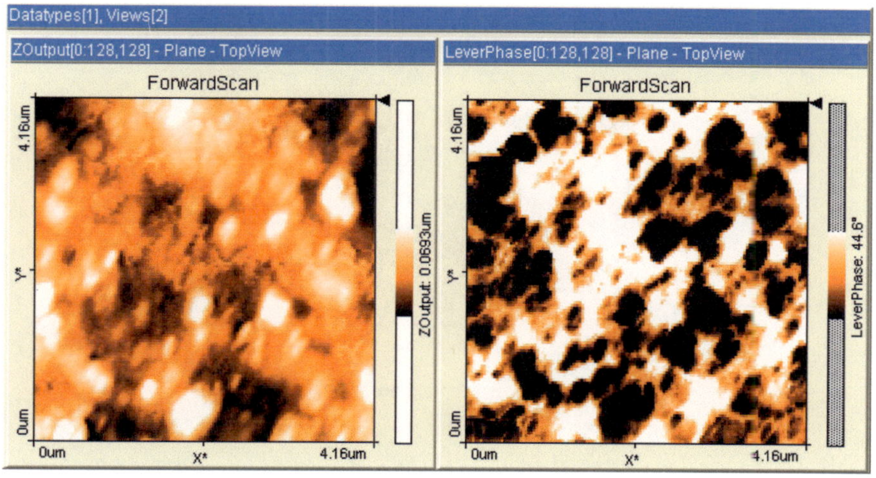

Figure 2.
SPM image of distribution in the rubber of milled nanoshungit. Scan 4.16 × 4.16 μm². Left—topography; and right—phase contrast.

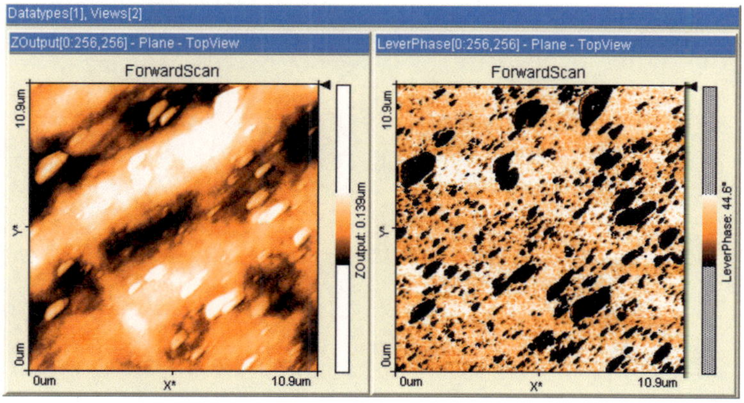

Figure 3.
SPM image of distribution in the rubber of milled nanoshungit, modified by organosilane Glymo. Scan 10.9 × 10.9 μm². Left—topography; and right—phase contrast.

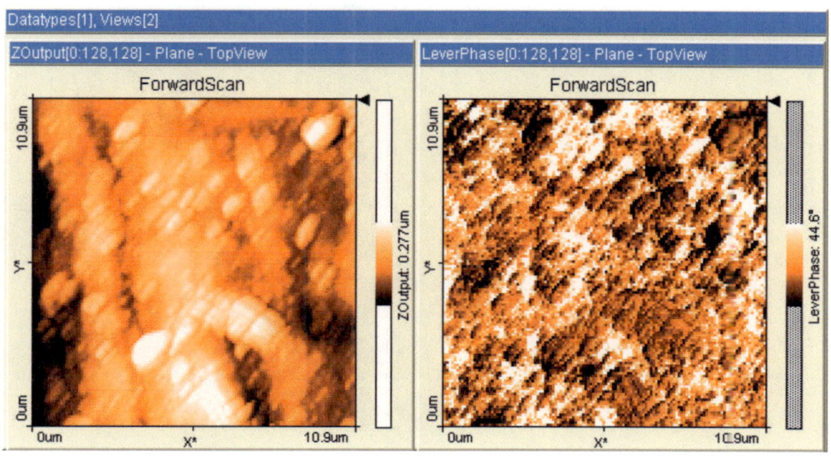

Figure 4.
SPM image of distribution in the rubber of milled nanoshungit, modified by organosilane thiol. Scan 10.9 × 10.9 μm². Left—topography; and right—phase contrast.

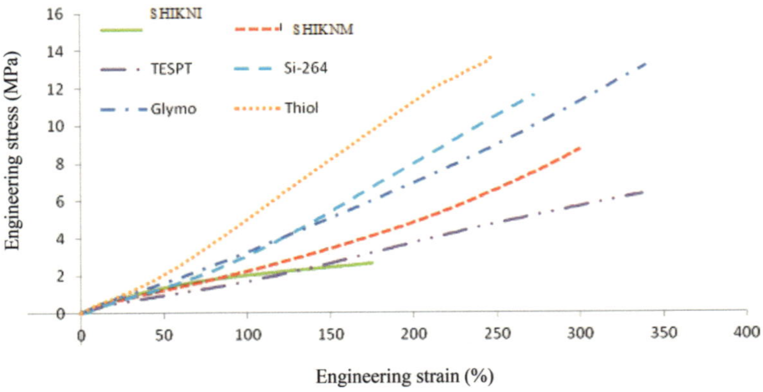

Figure 5.
The graphs of conventional strain-strength properties of studied composites with shungite. Engineering strain (%) are plotted in abscissa and engineering stress (MPa) in ordinates.

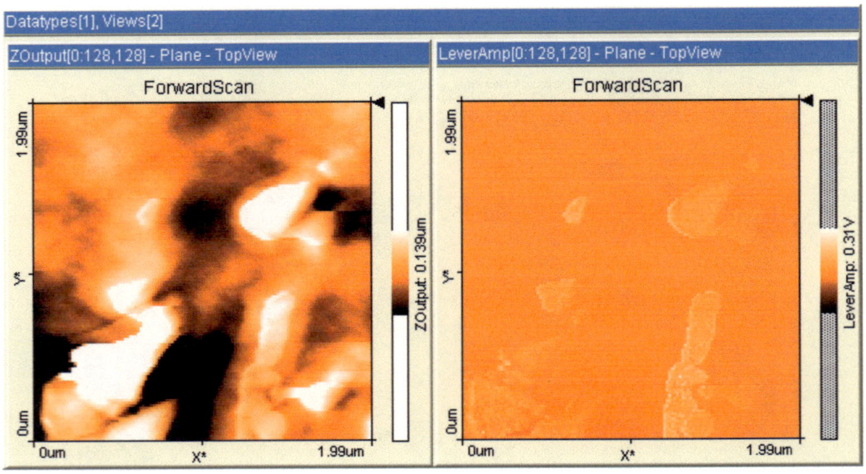

a)

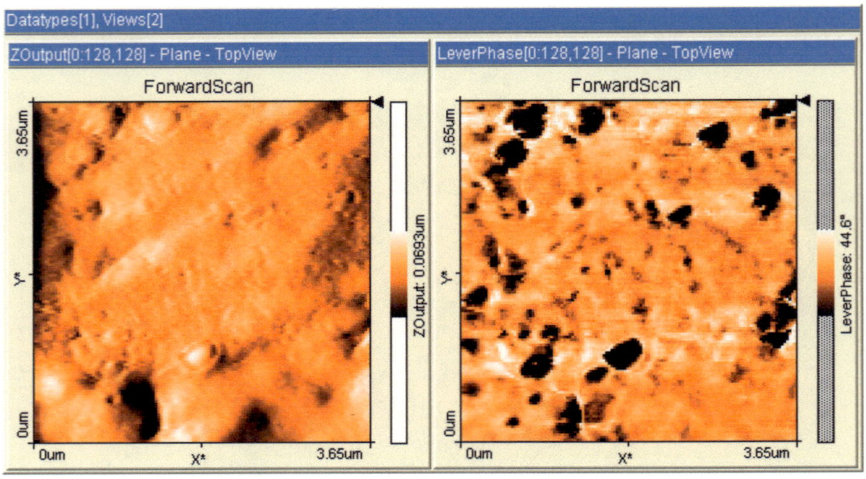

b)

Figure 6.
SPM images of the surface structure of rubber composites with (a) taurit (microstructural), scan 1.99 × 1.99 μm², and (b) taurit (nanostructured), scan 3.65 × 3.65 μm². Topography on the left, and material contrast on the right.

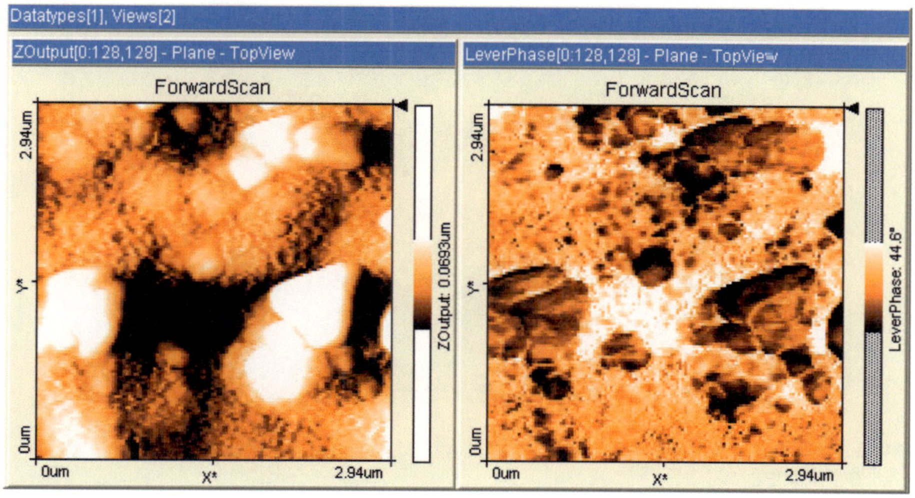

a)

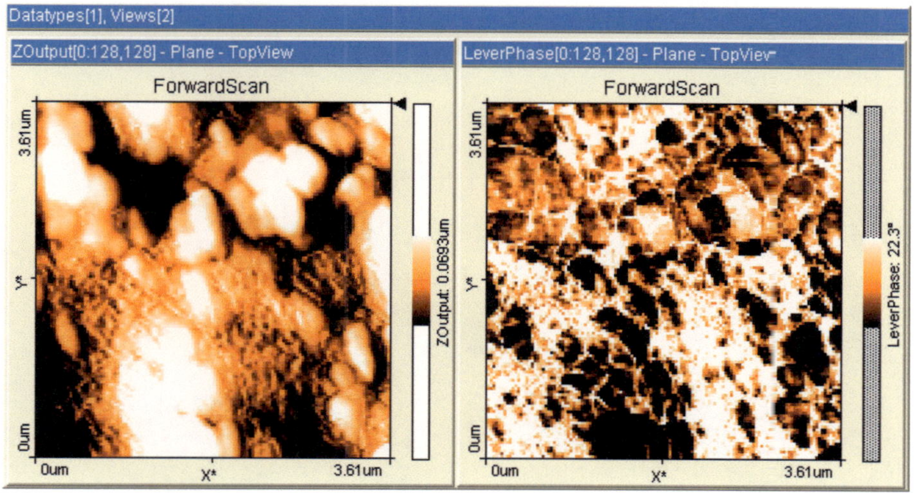

b)

Figure 7.
SPM images of the surface structure of rubber composites with (a) diatomite (microstructural), scan 2.94 × 2.94 μm², and (b) diatomite (nanostructured), scan 3.61 × 3.61 μm². Topography on the left, and material contrast on the right.

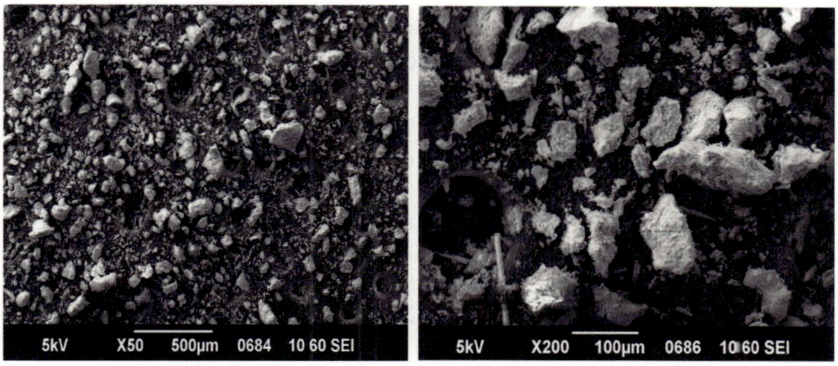

Figure 8.
SEM images of the original "Neosyl-120."

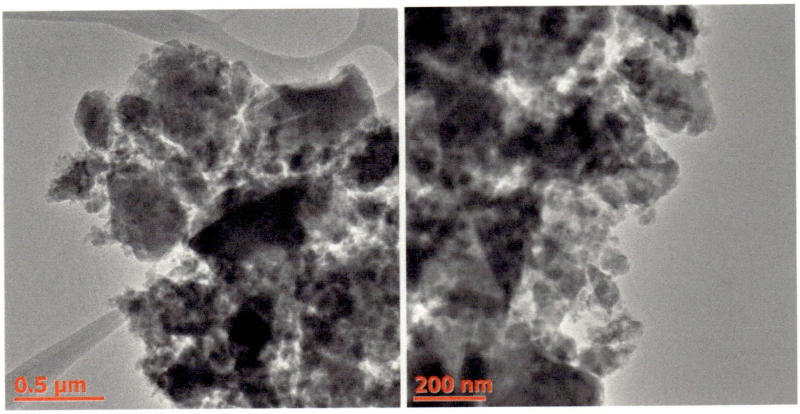

Figure 9.
TEM images of grounded Neosyl-120.

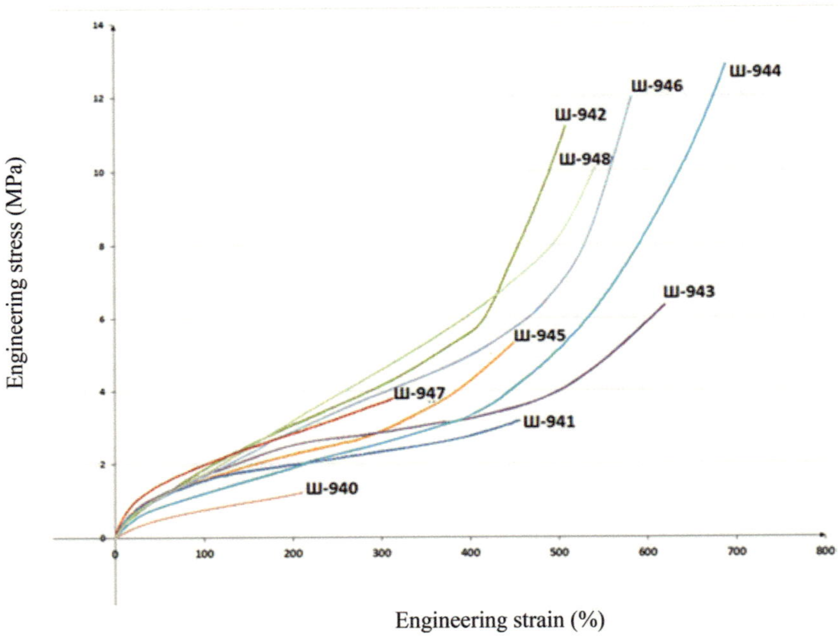

Figure 10.
The graphs of conventional strain-strength properties of studied composites with taurit, diatomite, and Neosyl-120. Engineering strain (%) are plotted in abscissa and engineering stress (MPa) in ordinates.

The graphs of conventional strain-strength properties of these elastomer composites with shungite [9]—are shown in **Figure 5**. These curves are typical ones for each type, corresponding to mean values from five repetitions for tests. Designations SHIKNI and SHIKNM in **Figure 5** mean rubber is filled by original and milled nanoshungit, respectively, and others by nanoshungit, modified by proper organosilanes.

SPM images of the surface structure of the synthesized composites with taurit and diatomit are shown in **Figures 6(a, b)** and 7**(a, b)**.

Scanning electron microscopy images of the original Neosyl-120 powder are presented in **Figure 8**. From the data obtained, it is clear that the particle size varies from submicrosized to large, with a size of the order of several hundred micrometers, with the second prevailing. Particles Neosyl-120 have a rough, sometimes porous, and layered structure, which implies a large value of its specific surface. However, the large value of the specific surface in this case cannot provide a high interaction area of the filler particles with the elastomeric matrix due to its high structurality.

Transmission electron microscopy has been used to obtain images of grounded Neosyl-120 particles in high resolution (**Figure 9**). Particle size has a wide distribution and amounts to tens and hundreds of nanometers. The particle morphology is close to spherical.

The physico-mechanical characteristics of the SBR-30ARK styrene-butadiene rubber-based vulcanizates, filled with taurit, diatomit, and Neosyl-120 of varying degree of dispersion (III-940–III-948), are presented in **Figure 10**.

4. Discussion of the results

Surface topography SPM images and the phase contrast of the composite material allowed to directly visualize the distribution of the fillers in the matrix rubber SBR-30ARK (**Figures 1–4**).

Analysis of **Figure 1** revealed the uneven distribution of original micro shungite filler in the rubber matrix. Predominant size of the shungite aggregates remains in the micron region. The distribution of aggregates and agglomerates of milled nanoshungit in the rubber (**Figure 2**) is considerably more homogeneous with a primary particle size of the filler already in the nanometer range. **Figure 3** shows that the use of organosilane Glymo as nanoshungit chemical modifier significantly improves the uniform distribution of the agglomerates and the aggregates in the matrix rubber. The aggregates and agglomerates of nanoshungit modified by organosilane Thiol (**Figure 4**) are distributed more uniformly at the rubber matrix even in comparison with modification by Glymo.

Analyses of the experimental data allow deducing certain conclusions. Modification of nanoshungit filler by organosilanes significantly improves the quality of rubber compounds. Using the Thiol organosilane, we obtained the highest tensile strength at 5 MPa more than rubber filled with nanoshungit without modification, with the extension not reaching 300%, which shows good mechanical properties of the vulcanizate. Organosilane Glymo sulfur-free showed an increase in rubber tensile strength, but its elongation was over 300%. The use of Si 264 as nanoshungit modifier also showed an increase in elastomer strength about 3 MPa relative to composites filled by nanoshungit without modification. This shows that the modification has been successfully completed. The sample with TESPT showed increased elastic-strength properties up to 30%.

Due to using force modulation or phase-contrast modes, separate aggregates and agglomerates of taurit or diatomit in sizes from 100 nm to 1 μm in the elastomeric matrix are visualized by SPM. The form of these aggregates and agglomerates in **Figures 6** and 7 is characterized by a sharp anisotropy of shape and an isolated heterogeneity of the filler particle structure. The images obtained show that in samples with microdispersed taurit or diatomit, the filler forms large inclusions with a size of several micrometers. When grinding the filler, a much more uniform distribution of nanostructured aggregates in the elastomeric matrix is observed. The results of experiments demonstrate also that the dispersity of the filler has a significant effect on tensile behavior of composites. Materials with microdispersed fillers have much lower tensile strength than samples with nanodispersed fillers. When grinding taurit to the nanodispersed phase, the strength of the vulcanizates increased from 3 to 12 MPa, and diatomit from 5 to 13 MPa, respectively. Grinding of particles of the Neosyl-120 products allows increasing the level of interaction of the filler with the elastomeric matrix, which leads to a noticeable improvement in the elastic-strength properties: the conventional strength increases from 3 to 10 MPa. Grinding of these fillers did not lead to significant changes in the modulus of elasticity.

Theoretical analysis of obtained results for these different fillers in such elastomer matrix allows to conclude that the reason for strength reinforcement of these composites is due to the redistribution of valence π electrons of butadiene-styrene rubber between acceptor fillers and the donor matrix of the elastomer. The resulting interatomic interactions lead to more uniform distribution of fillers in the rubber matrix and the corresponding effects of enhancing the physico-mechanical characteristics of the composites.

5. Conclusions

The comprehensive study of SPM and physico-mechanical tests of the new model samples of rubbers based on butadiene-styrene rubber SBR-30 ARK filled with micro- and nanoshungit modified by organosilanes yielded important data on the properties of the investigated objects. Comparison of the SPM data with the results of tensile testing machine on the characteristics of these composites has allowed to conduct cross-correlation. The use of SPM made also possible to determine the patterns of distribution of aggregates and agglomerates of taurit, diatomite, and Neosyl-120 in composites with butadiene-styrene rubber. The morphology and micro-nanometer range of the length of these aggregates in the synthesized materials has been established. A correlation between the enhancement of the physico-mechanical properties of the synthesized composite materials and the distribution of the used fillers in the elastomeric matrix was established. The proposed method of grinding the studied minerals in the laboratory allows to significantly increase the strength characteristics of the rubbers filled by them, indicating that they can be used as reinforcing fillers for elastomeric composites. These results are important for both the fundamentals of the theory of the reinforcement of composites and practical applications in industry.

Acknowledgements

The work was done on the state assignment at the Institute of Applied Mechanics of the Russian Academy of Sciences for the period 2019–2021. State registration number: AAAA-A19-119012290177-0.

Author details

Hammat H. Valiev*, Alexander N. Vlasov, Yury V. Kornev, Yuliya N. Karnet, Nikolay A. Semenov and Oleg B. Yumaschev
FBGUN Institute of Applied Mechanics of the Russian Academy of Sciences, Moscow, Russia

*Address all correspondence to: hhvlv@mail.ru

References

[1] Yanovsky YG, Obraztsov IF. Some aspects of computer modeling of the structure and micromechanical properties of promising polymer composite materials. Physical Mesomechanics. 1998;**1**:135-142

[2] Yanovsky YG. Nanomechanics and Strength of Composite Materials. Moscow: IPRIM RAS; 2008. 179 p

[3] Vlasov AN, Volkov-Bogorodsky DB, Yanovsky YG. The influence of the size of inclusions on the mechanical properties of elastomeric composites. Reports of the Academy of Sciences (RAS). 2012;**443**:435-437

[4] Gamlitsky YA. Nanomechanics of the phenomenon of reinforcement of filled elastomers. International Journal of Nanomechanics Science and Technology. 2013;**4**:179-196

[5] Musina US. Koksu shungite rocks in the process of ensuring ecological balance. Bulletin of the St. Petersburg State Institute of Technology. 2014;**23**(49):79-82

[6] Demidov IN, Shelekhova TS. Diatomites of Karelia (Features of Formation, Distribution, Use Prospects). Petrozavodsk: Karelian Research Center of RAS. Institute of Geology; 2006. 89 p

[7] Valiev HH, Vorobyev VV, Karnet YN, Kornev YV, Yumashev OB. Application of quantum-chemical modeling results in experimental investigations of silicone composites. Materials Physics and Mechanics. 2017;**32**:293-297

[8] Rubber. Method of the Determination Elastic and Tensile Stress-Strain Properties [Internet]. 1978. Available from: https://znaytovar.ru/gost/2/GOST_27075_Rezina_Metod_oprede.html

[9] Valiev HH, Karnet YN, Kochurov NL, et al. Atomic force microscopy and physical - mechanical properties of new elastomer composites. Materials Physics and Mechanics. 2016;**26**:45-48

Chapter 4

Methodologies for Achieving 1D ZnO Nanostructures Potential for Solar Cells

Yeeli Kelvii Kwok

Abstract

One-dimensional (1D) nanostructures are generally used to describe large aspect ratio rods, wires, belts, and tubes. The 1D ZnO nanostructures have become the focus of research owing to its unique physical and technological significance in fabricating nanoscale devices. When the radial dimension of the 1D ZnO nanostructures decreases to some lengths (e.g., the light wavelength, the mean of the free path of the phonon, Bohr radius, etc.), the effect of the quantum mechanics is definitely crucial. With the large surface-to-volume ratio and the confinement of two dimensions, 1D ZnO nanostructures possess the captivating electronic, magnetic, and optical properties. Furthermore, 1D ZnO nanostructure's large aspect ratio, an ideal candidate for the energy transport material, can conduct the quantum particles (photons, phonons, electrons) to improve the relevant technique applications. To date, many methods have been developed to synthesize 1D ZnO nanostructures. Therefore, methodologies for achieving 1D ZnO nanostructures are expressed, and the relevant potential application for solar cells are also present to highlight the attractive property of 1D ZnO nanostructures.

Keywords: ZnO, one dimensional, nanostructures, chemical vapor transport and condensation (CVTC), chemical vapor deposition (CVD), metal-organic chemical vapor deposition (MOCVD), vapor-liquid-solid (VLS), hydrothermal, electrochemical, solar cell

1. Introduction

To date, nanotechnology is the operation of matter on an atomic and molecular scale. Generally speaking, the size scale in nanotechnology including materials, devices, and other structures is at least from 1 to 100 nanometers in one dimension. The revolution of the nanotechnology is taking a crucial effect on the different fields, such as commercial sectors, engineering, science, drug delivery, sensors, and the construction industry. Nanostructures in such size have made steadily increasing attraction because of their attractive and captivating properties, same as their fascinating applications complementary to the materials in bulk. The interesting properties of materials in nanoscale (both physical and chemical) can make the efficacy enhanced distinctively in mechanical strength, (photo)catalysis, optical sensitivity, and (thermal and electrical) conductivity which enable applications such as improved materials with higher properties, storage devices of the electronic and energy, sensors, and catalysts [1–8].

According to the dimensions in nanometer scale size, nanostructures can be classified into the following three groups:

a. Zero-dimensional (0D) nanostructures (quantum dots, nanoparticles, or nanoclusters)

b. One-dimensional (1D) nanostructures (nanowires, nanorods, nanotubes, nanoribbons, nanobelts, or nanocables)

c. Two-dimensional (2D) nanostructures (super-thin films, multiple-layer films, or superlattices)

In comparison with 0D nanostructures, it is easier to investigate the relationship between the mechanical properties, optical and electronic transport, and the confinement of the size and dimensionality for 1D nanostructures. Moreover, 1D semiconductor nanomaterials have an extremely crucial effect of the active components and interconnect in the nanoscale electronic and photonic devices fabrications.

Up to now, 1D nanostructure is generally used to describe large aspect ratio rods, wires, tubes, and belt and tubes and has been the key point of investigation due to its attractive physical and technological significance in fabricating nanoscale devices. When the radial diameter of the 1D nanostructure is lower than some lengths (the path of the phonon mean free, the light of wavelength, Bohr radius, etc.), the effect of the quantum mechanics will be crucial. Owing to the large surface-to-volume ratio and the confinement of two dimensions, nanowires possess the definitely attractive electronic, magnetic, and optical properties. In addition, because nanowires' aspect ratio is extremely large, the quantum particles (photons, phonons, electrons, etc.) can be conducted directly easily to make the nanowires as the ideal candidate for the energy transport materials to enhance the relevant technique applications [9–17].

Today, numerous approaches have been researched to synthesize 1D nanostructures. Two fundamental steps are essentially involved in the evolution: nucleation and growth. A lot of solid materials with 1D nanostructures in nature are controlled by the bonding in the structure of crystallography in the highly anisotropic. The materials need common growth conditions including chemical vapor deposition (CVD), wet-chemical routs, and template-assistant methods. In classification, all the contemporary approaches are divided into bottom-up and top-down methods. The most important issue for developing a new synthetic method is to control the dimensions, morphology, and uniformity of nanostructures. When making a method to synthesize the nanostructures by the synthetic effects, it is definitely important to control the related morphology (or shape), dimensions, and uniformity simultaneously. To obtain 1D growth nanostructures, several chemical methods were generated. The current common six different strategies are (1) the reduction of a 1D microstructure in size, (2) 0D nanostructure self-assembly, (3) by a capping reagent kinetic control, (4) a template usage for direction, (5) a liquid droplet confinement as in the vapor-liquid-solid process, and (6) the control of a solid with the anisotropic crystallographic structure.

Generally, there are four popular mechanisms for understanding the synthesis of 1D nanostructure materials. They are the mechanism of vapor-liquid-solid (VLS), the mechanism of oxide-assisted growth (OAG), the mechanism of vapor-solid (VS), and the mechanism of solution-liquid-solid (SLS).

1.1 Vapor-liquid-solid (VLS)

In the VLS mechanism, a liquid metal cluster or catalyst, such as Au, Fe, Ni, or Co, is taken as the energetically favorable point of the gas-phase reactant absorption.

The reactant supersaturates and segregates from the cluster and then grows into a 1D structure of the material, the diameter of which is limited by the size of the liquid metal catalyst that can be achieved under equilibrium conditions [18–21].

1.2 Oxide-assisted growth (OAG)

This kind of the synthesized technique, in which oxides replaced by metals have a crucial effect on the nucleation induction and the nanowire growth, can produce the high-purity 1D nanomaterials in the large quantities, and the metal catalysts do not need any more. The 1D nanomaterial synthesis with the mechanism of the oxide-assisted growth is the extension of the traditional vapor-liquid-solid method with the metal catalyst. Moreover, it can be taken to make the nanowires by other materials than silicon [22–26].

1.3 Vapor-solid (VS)

For the mechanism of the vapor-solid (VS), the size of the nucleation site is critical for defining the rod diameter when the vapor supersaturation is appropriately controlled. Metal catalysts are not necessary. Three stages can be summarized as the illustration: (i) The source forms vapor phase, (ii) the vapor is transported by the carrier gas and deposits on the substrate to form crystalline nuclei, and (iii) the defects of the nuclei become the growth points, and the reactive vapor molecules further grow into nanostructures [27, 28].

1.4 Solution-liquid-solid (SLS)

Solution-liquid-solid (SLS) phases are involved in the nanowire growth that is in fact an analogy to the conventional whisker growth via vapor-liquid-solid (VLS) mechanism. The difference is that the vapor phase involved in the VLS growth is now substituted by a solution phase in the SLS mechanism. In turns out, however, the nanowires prepared by the SLS mechanism have a varying diameter ranging from 10 to 150 nm, which is not uniform [29, 30].

2. Property of one-dimensional (1D) nanomaterials

Up to now, it is still a challenge to accurately characterize the property of 1D nanostructures due to the constrains of the current measuring techniques, such as (1) the size of 1D nanostructures is too small to adopt the well-established testing techniques and (2) 1D nanostructures different from bulk materials are hard to pinpoint at the desired location. Therefore, the relevant techniques should be explored to detect the property of 1D nanostructures accurately.

2.1 Mechanical property

1D nanomaterials behave qualitatively different from the conventional bulk materials when the size reduces to nanoscale. It is well known that the increments of yield stress and the hardness of a polycrystalline material are consistent with the decrement of the grain size to the micrometer scale, and this significant phenomenon is defined as the Hall-Petch effect. For the single-crystalline 1D nanostructures, their property is extremely higher than that of the counterparts in the larger dimensions.

2.2 Thermal property

For 1D nanostructures, the great reduction of the melting point of a solid material is very obvious. Because of the special characteristics of nanocrystal materials, the high specific boundary area means large stored interface energy. Consequently, 1D nanostructures can be tailored precisely on the basis of their thermal property, such as synthesizing and annealing temperatures [31–33].

2.3 Electronic property

For 1D nanostructures, the grain size and boundaries are the dominated factors effected on the electron mean free path and resistance. There is a trend that many physical properties of 1D nanostructures like optical, magnetic, and electrical properties will be enhanced distinctively when the size or dimension of the material reduces to nanosize (~10^{-9} m) scale. Moreover, such 1D nanostructures (such as nanowires, nanorods, etc.) are generally prone to be enriched with many surface defects and oxygen and cation vacancies due to their low formation energy within nanoscale materials [34, 35].

2.4 Magnetic property

Magnetic properties of materials are fundamentally determined by the magnetic couplings at the atomic level. Unlike bulk ferromagnetic materials, which usually form multiple magnetic domains, 1D nanomaterials consist of the simple magnetic domain resulted in obvious difference in several important aspects from the property of their bulk counterparts [36–40].

2.5 Optical property

The confinement of the size has a significant effect on the energy levels of the nanowire determination when its diameter decreases to some critical length (Bohr radius). Results indicate that the nanowire absorption edge of silicon is obviously blue-shifted because the bulk silicon indirect bandgap is only 1.1 eV. The characteristics of the absorption spectra are sharp and discrete along with the photoluminescence (PL) in the relatively strong "band-edge." At the same time, along the longitudinal axes of the nanowires, the emitted light is highly polarized [41–44].

3. Zinc oxide

Zinc oxide (ZnO) has been investigated for a long time as it is an amazing material with multiple functions. ZnO is a direct wide bandgap semiconductor material with piezoelectric and photoelectric properties. ZnO has a wide direct bandgap of 3.37 eV which is similar to GaN and a high exciton binding energy of 60 meV at room temperature. The wide bandgap gives good optical transparency to visible light which makes ZnO a suitable candidate for short wavelength photonic applications (UV and blue spectral range). ZnO has a non-central symmetric wurtzite structure, and the relevant hexagonal unit cell (a = 3.25 Å, c = 5.20 Å) packed O^{2-} closely and stacked Zn^{2+} layers alternately along the c-axis direction. Due to the unique fascinating property in electronics, optics, photonics, and magnetics, ZnO provides an impact on applications in various areas, such as solar cells, supercapacitors, sensors, catalysis, light-emitting, actuators, and biomedical devices. ZnO has equal importance in relation to silicon-based 1D nanostructures in the field of 1D nanostructures, and it has an increasing influence in developing nanotechnology. To date, various quasi-one-dimensional nanostructures of ZnO have been synthesized, i.e., nanowires, nanobelts, and nanotubes [45–47]. **Figure 1(a–c)** expresses the images of ZnO rods taken by SEM synthesized at pH 8. It indicates that the length

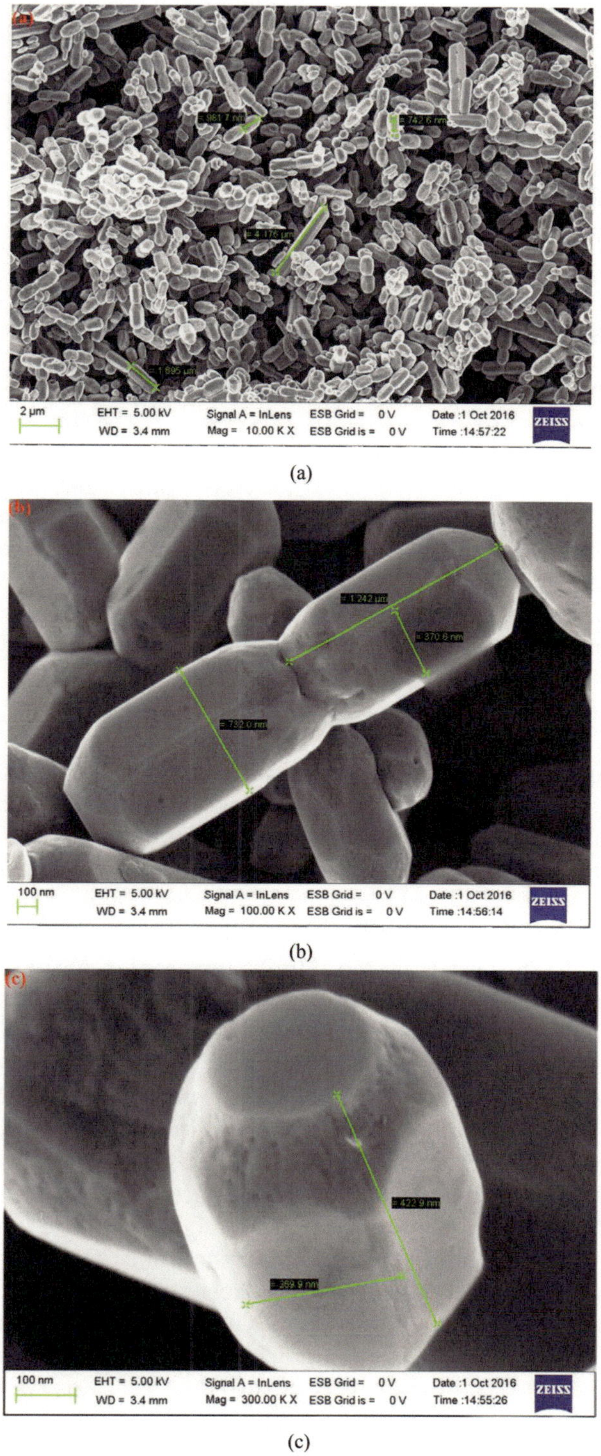

Figure 1.
(a–c) ZnO nanorod images taken by SEM synthesized at pH 8 (a) 10 k×, (b) 100 k×, and (c) 300 k×.

of synthesized ZnO nanorods is about 4 μm with the diameter of around 700 nm and ZnO nanorods with the flat top surface, and they stack one by one through polar surfaces. From the crystal structure of the ZnO, the ions of Zn and O are arranged alternatively through *c* axis where the bottom surface is O^{2-} terminated (000-1) and the top surface is Zn^{2+} terminated (0001). The surfaces of the flat top explored in the nanorods of ZnO are contributed to the polar surface disappearance. In the basis solution with the weak volume, the precipitate of $Zn(OH)_2$ solid exists in the reactant solution. Owing to the dipole interaction, $Zn(OH)_2$ solid is taken as the polar surface that could easily make the positive and negative surfaces of ZnO crystal incorporate efficiently. Therefore, the surface energy of the polar surfaces is relatively high than that of the nonpolar surfaces, disappears at the first when the nonpolar surfaces start to slowly grow, and appears in the last stage of ZnO nanostructure crystal growth [48].

4. Methodologies for achieving 1D ZnO nanostructures

ZnO nanostructures have already been synthesized by various methods. There are mainly two methods to prepare 1D ZnO according to the state-of-growth medium: vapor phase process and liquid phase process. The vapor phase method includes chemical vapor deposition (CVD), metal-organic chemical vapor deposition (MOCVD), molecular beam epitaxy (MBE), pulsed laser deposition (PLD), etc. The liquid phase method includes hydrothermal method, electrochemical method, sol-gel method, and so on. The growth mechanism can be classified into mainly three categories: vapor-solid (VS) growth, vapor-liquid-solid (VLS) growth, and solution-liquid-solid (SLS) growth.

4.1 Chemical vapor transport and condensation (CVTC)

The chemical vapor transport and condensation (CVTC) method is processed in a tube furnace. The substrates coat with a layer of Au thin film with thermal evaporation methods to control the thickness of the film in the CVTC system. ZnO powder and graphite powder in the same amounts are mixed together and milled effectively; then put them into a boat made by alumina. Then, the alumina boat and the substrate with Au film are placed into a small quartz tube. Usually, the substrates are kept 5–10 cm away from the alumina boat center. At the furnace center, the boat of the alumina sits, and at the downstream of the argon flow, the substrates are positioned. The system is raised to 800–900°C and maintained for 5–30 min. Light or dark gray materials are attained on the surface of the substrate.

The transmission electron microscope (TEM) image of an alloy tip on a thin nanowire is shown in **Figure 2**. According to the features of the alloy tip, it indicates a clear growth procedure by vapor-liquid-solid (VLS) growth mechanism. The synthesized tips of nanowires shown in **Figure 2a** are grown by both the hydrogen and graphite reduction approaches. The relevant high-resolution transmission electron microscope (HRTEM) image of a single-crystalline ZnO nanowire is shown in **Figure 2b**. It expresses that the space of the adjacent lattice planes of 2.56 ± 0.05 Å is consistent to the distance of two crystal planes in (002). Consequently, it indicates that the growth direction for the ZnO nanowires is <001>. Meanwhile, it is also observed in the high diffraction intensity of (001) peaks of XRD results to confirm that the preferential growth direction is <001> [49, 50].

Figure 3 expresses a representative image of a patterned ZnO nanowire synthesized on silicon (Si) substrate with the patterned Au islands taken by scanning electron microscopy (SEM). The diameters of the nanowires are from 20 to 120 nm,

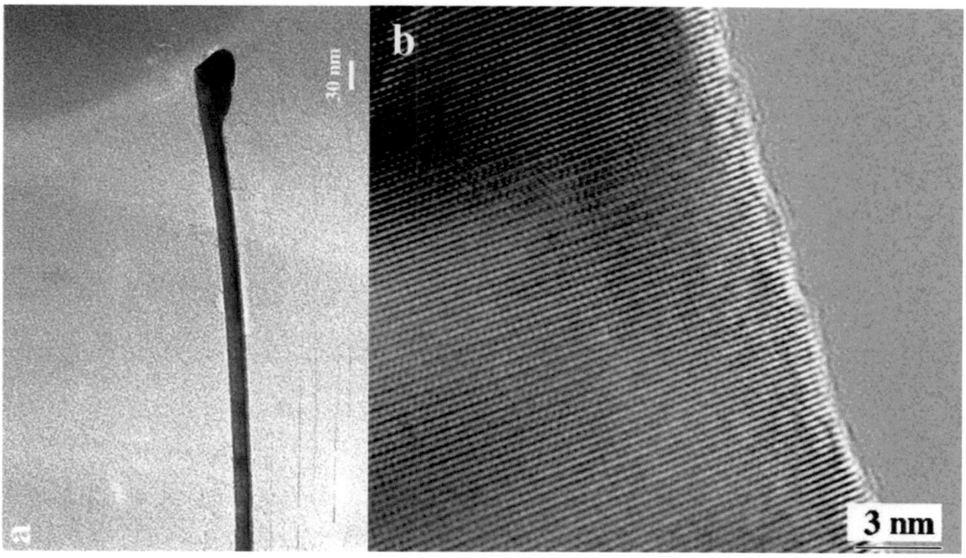

Figure 2.
(a) A thin ZnO nanowire image taken by TEM with a Zn/Au alloy tip. (b) the lattice fringes of high-resolution TEM image of a single-crystalline ZnO nanowire.

and the length is 5–20 μm. The directions of the grown nanowires on (100) Si substrate are random.

SEM results show that the flexible, long, fine ZnO nanowires grow extensively from the hexagons' edges. The growth of nanowires is consistent with the copper grid in the hexagonal pattern abundantly. Interestingly, a complicated intricate network is formed due to lots of the nanowires connecting with the neighboring metal hexagons. SEM images for the higher magnification of the same sample illustrate more detailed particulars as shown in **Figure 3b**.

ZnO NWs grown under the vapor-liquid-solid process begin together with the reductive Zn gaseous reactants dissolution into the Au catalyst liquid droplets in nano-size, and then the alloy metal is formed followed by the supersaturation of Zn along with the single-crystalline wire growth. The schematic growth mechanism is expressed in **Figure 4**. In this method, the diameter, density, and location of ZnO NWs can be controlled according to the desired characteristics. As a result, ZnO NWs with the required properties can be attained and tailed successfully. However, the metal catalyst affects the purity of the product which can lower the performance of the nanowires [51–56].

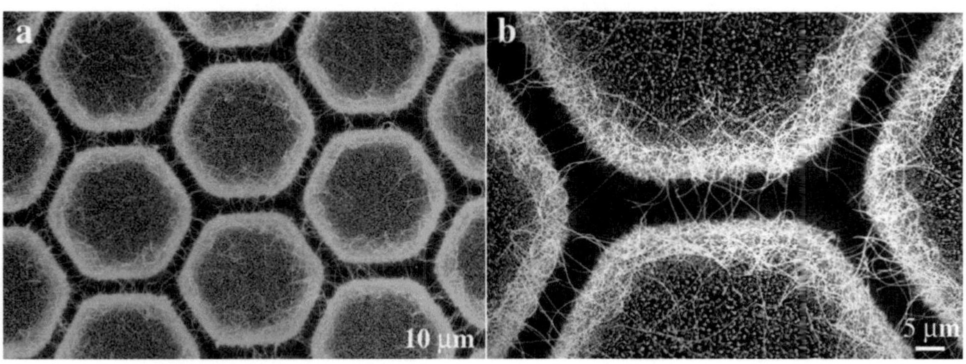

Figure 3.
(a) From the patterned Au islands images of ZnO nanowire networks synthesized taken by SEM. (b) Higher magnification SEM image for the same sample.

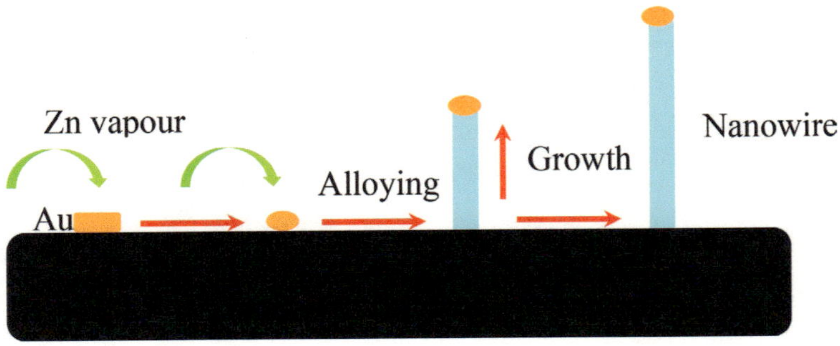

Figure 4.
Schematic mechanism of ZnO NW growth.

4.2 Chemical vapor deposition (CVD)

The aligned 1D ZnO NWs fabricated by patterning metal catalytic particles through VLS growth involve tedious lift-off processes for patterning metal catalysts and may lead to serious contamination in complementary metal oxide semiconductor processing. To avoid the effects of the catalyst, CVD without the catalyst has been developed. The zinc oxide NW growth is proceeded and developed earlier when the flow-type reactor is at a reduced pressure of the elemental vapor phase synthesis. Highly purified metallic granulated Zn (99.99%) is placed in the boat made by the alumina. After that, it is inserted at a quartz ampoule end which is sealed at one end. Meantime, a wide slit in the ampoule is at the open end. The substrates are mounted with their front sides up and below the ampoule. Then, a horizontal two-zone flow-type quartz reactor is used for putting the ampoule inside in order to make sure that the source of zinc is situated in one of the zones (evaporation zone), the substrates, and in the other (growth zone). The arrays of ZnO growth are very allergic to the processing parameters, especially for the well-aligned nanorods. Wang ZL and his co-workers firstly fabricated a ZnO nanobelt in 2001 using catalyst-free CVD methods and applied it to the devices [57]. Furthermore, Wang's group also achieved the formation process of a rectangular cross section by simply evaporating ZnO powder at elevated temperatures. The structures of as-synthesized nanobelts are uniform. Most of the nanobelts are single crystals and free from defects and dislocations. To obtain the ZnO nanowire arrays (NWAs) via catalyst-free CVD methods, many groups have adopted different methods to reach the objective [58–62]. One method is to adopt a ZnO film layer to induce the growth of ZnO NWs. The nucleation stage of ZnO nanocrystals plays an important role. Firstly, the zinc metal droplet condenses on the surface of the substrate due to the different heating temperatures in the evaporation and growth zones. The diameter of catalyst droplets can be controlled by zinc and oxygen partial pressure which is different from the Au catalyst growth mechanism. In addition, Menzel A et al. deeply investigated the method for tuning the growth mechanism of ZnO nanowires under the various conditions and put out the related parameters for a controlled NW growth by CVD method to change the nanowire shapes. The results are shown in **Figure 5** [63].

4.3 Metal-Organic chemical vapor deposition (MOCVD)

The MOCVD method, apart from its increasing advantages because of its unique characters of the industry, has been illustrated to be effectively taken to ZnO NW synthesis with good controllable shape, high quality, and reproducibility. The growth of ZnO NWs with high quality by catalyst-free MOCVD was explored firstly by Park et al. [64]. After that, the huge effort has been put into the field of

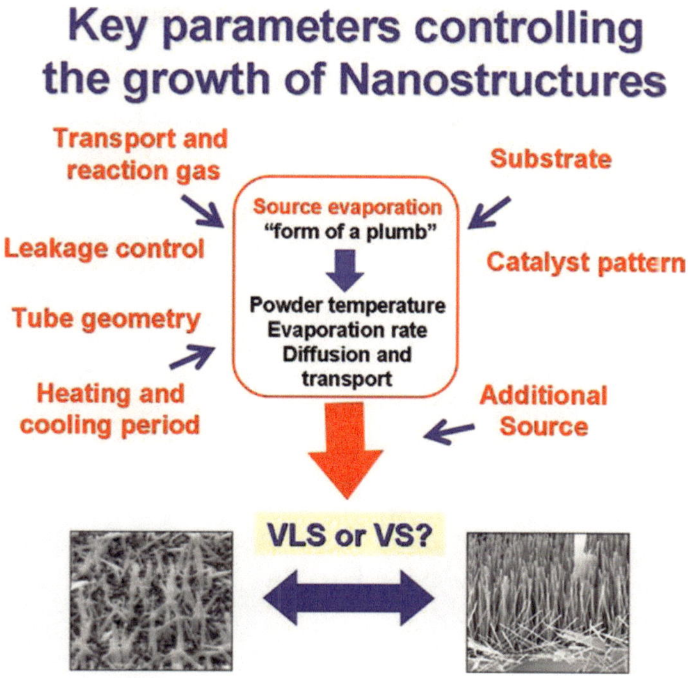

Figure 5.
Schematic diagram of the key parameters controlling the NW growth.

the synthesis of the ZnO nanostructures, and great progress has subsequently been obtained [65–69]. Up to now, the contributions relevant to MOCVD growth of ZnO nanostructures, diethylzinc (DEZn), dimethylzinc (DMZn) as zinc precursor, nitrogen or argon as carrier gas, and low reactor-pressures have been mostly used. Park's group reported the growth of ZnO NWs, which requires no metal catalysts. The nanorod growth temperature was as low as 400°C. However, the preparation machine and source materials are more expensive than other methods which have hindered its practical applications.

4.4 Chemical solution method

Chemical solution deposition is one of the commonly employed synthesis methods for ZnO nanostructures, particularly in large-scale fabrication for device purposes [70–75]. Chemical reactions between different precursors play a key role in the synthesis. The advantages of the solution method include the economical synthesis, the large scale, and the low temperature. For ZnO nanomaterials growing on a substrate, the approaches of solution mostly adopt the hydrothermal procedure by a kind of solution in an aqueous which includes an organic amine and zinc salt. In addition, in order to enhance the ZnO NR alignment on the substrate, a textured ZnO nanocrystal or a ZnO thin film is taken as a seed layer.

In a typical chemical solution synthesis, a layer of ZnO seed layer is spread over a Si substrate by dipping or sputtering. This kind of seeding method is simply suitable for different substrates. The ZnO seed layer thickness is usually 10–200 nm. A calculated amount of zinc nitrate hexahydrate is dissolved in 80 mL deionized water to obtain 1–40 mM solutions for growth solution preparation. Then, the pH of the solution for ZnO growth is adjusted by ammonia water addition. The amount of the ammonia water addition related to the target pH and the zinc salt concentration is usually about 0.1–5 mL. The growth of the hydrothermal ZnO NRs is explored with Zn/Si substrate suspended upside down in a kind of Teflon-capped glass bottle which is full of the growth solution. The temperature of ZnO growth ranges from 60 to 90°C and the reacting time is 6 h. When the synthesis is finished, the substrate is taken from the reactant solution. At the same time, the substrate is rinsed by the DI water and dried successively. Therefore, the morphology (length, diameter) of the synthesized nanorods relied on the relevant parameters, for instance, zinc seed layer morphology, pH, growing temperature, and zinc salt concentration.

Figure 6a shows ZnO nanorods viewed normal to the surface grown by the two-step chemical bath deposition process, in which the facets are exactly crystalline hexagonal and its average ratio of the aspect is about 3 ± 1 [76]. **Figure 6c** illustrates the sulfidation of ZnO nanorods with definitely well particle decoration of the entire surface of ZnO nanorods, and the end and side of facets are not in the well-defined morphology. With the reaction increasing, a more uniform film coated

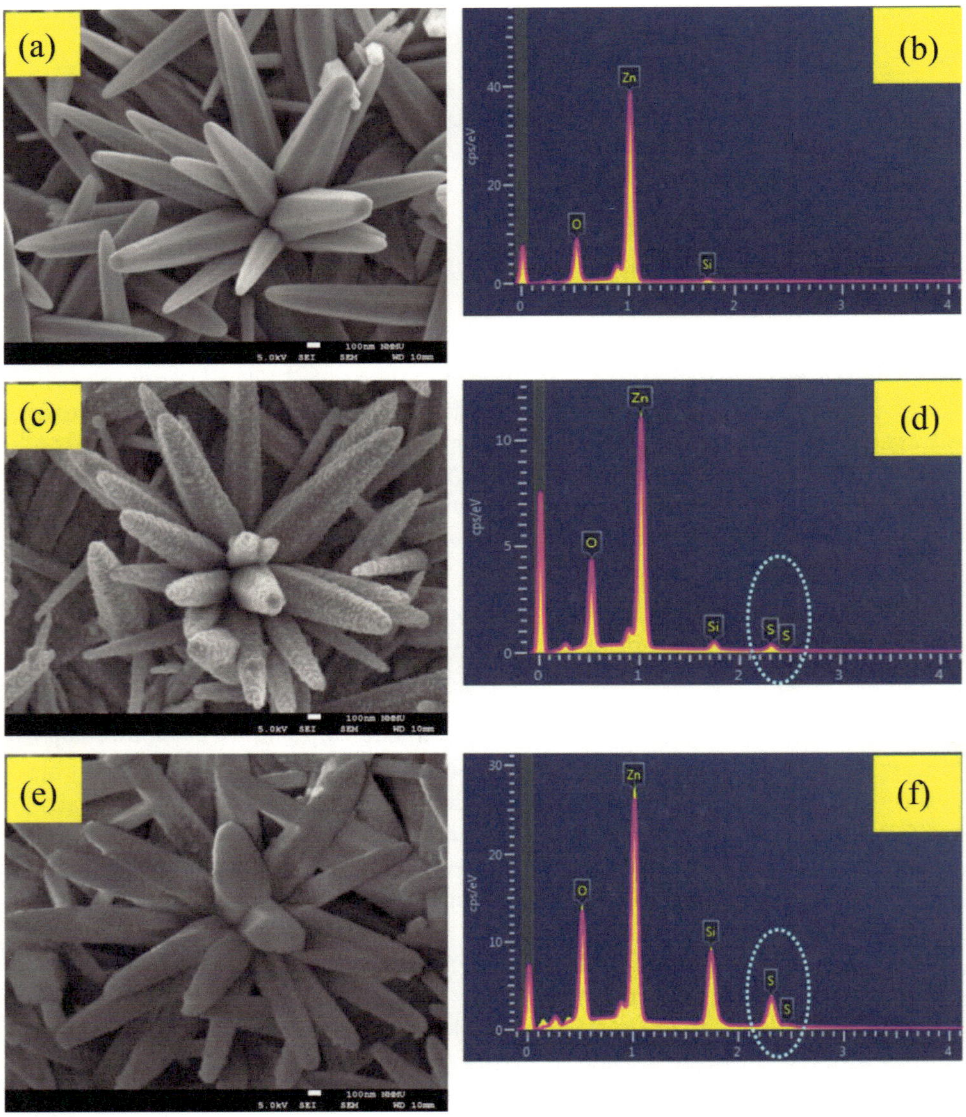

Figure 6.
ZnO nanorod (a) initial morphology (sample A) before sulfidation, (c) sulfidation for 90 min at 75°C in 160 mmol $Na_2S_{(aq)}$ *(sample B), and (e) treatment further for 90 min at 75°C in 160 mmol* $Zn(NO_3)_2{\cdot}6H_2O_{(aq)}$ *(sample C). EDS image (b) results of a, (d) results of c, and (f) results of e.*

on ZnO nanorods can be easily observed as shown in **Figure 6e**. Moreover, the side and end facets of the synthesized ZnO nanorods as shown in **Figure 6e** become more smooth (cf. **Figure 6a**, **Figure 6c**, and **Figure 6e**). Meanwhile, EDX results agree well with the relevant SEM results.

The related TEM results are shown in **Figure 7**. It expresses that ZnO nanorod is coated with an uneven film. The selected area diffraction (SAD) result as shown in

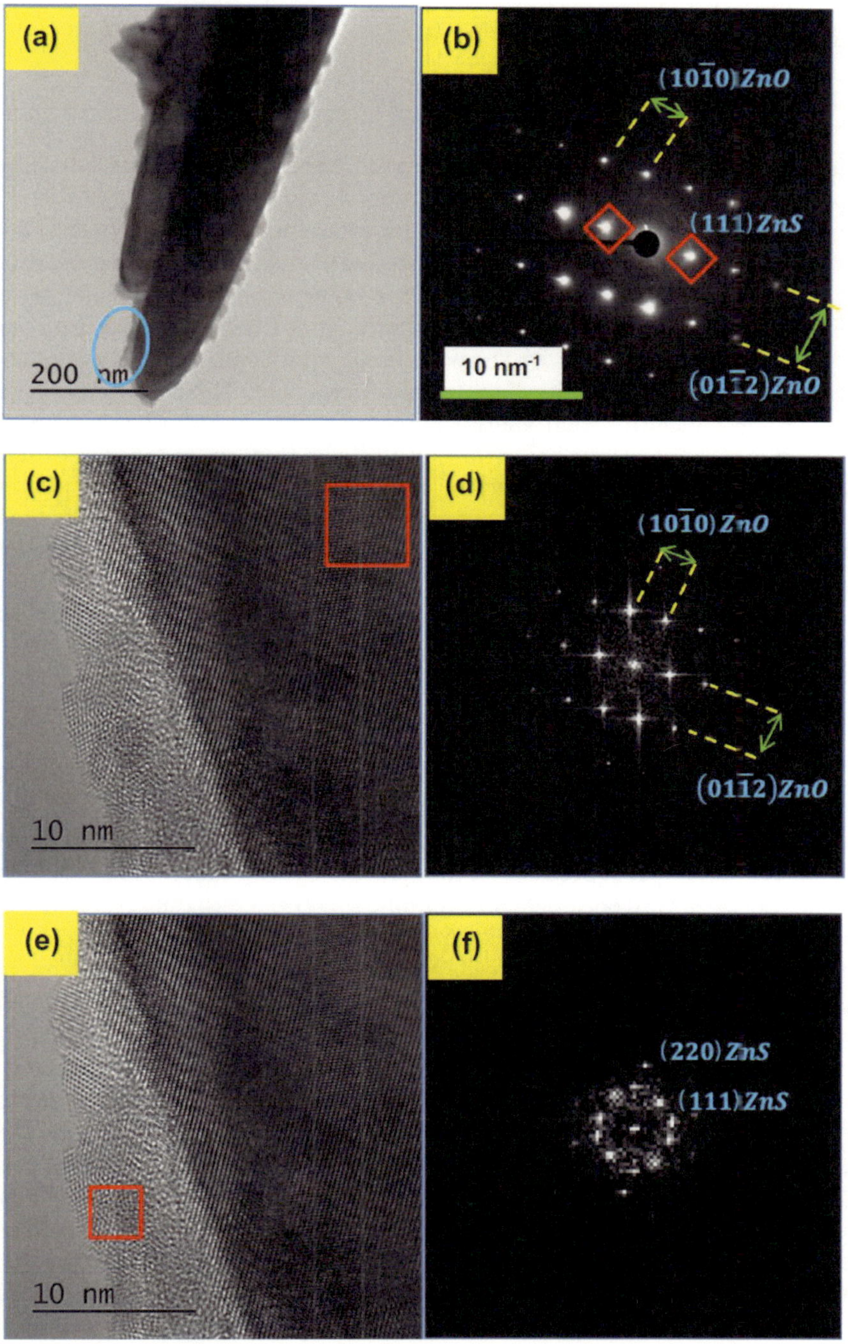

Figure 7.
*Sample B images of HRTEM. (a) Sulfidation nanorod (bright field), (b) region in blue ellipse (**Figure 7a**) (selected area diffraction (SAD) pattern), (c and e) interface of ZnO-ZnS (high-resolution images), (d) region in red (**Figure 7c**) live Fourier transformation, and (f) region in red (**Figure 7e**) live Fourier transformation.*

Figure 7b assures the crystalline structure in which the bright spots are the crystalline ZnO and the rings are the polycrystalline ZnS. The interplanar distances related to ZnO (01-12) and (10-10) and ZnS (111), together with the relationship of the partial epitaxial between ZnS shell and ZnO core where (10-10) ZnO//(111) ZnS, are confirmed. Furthermore, the calculated parameters of the lattice of ZnO core are 5.35 ± 0.01 Å at the *c*-axis and 3.29 ± 0.01 Å at the *a*-axis.

5. Solar cells

5.1 Advantages of ZnO NWs/NR arrays

ZnO has a large bandgap (3.37 eV) *n*-type semiconductor which can be easily synthesized into large-scale arrayed 1D ZnO structures and the patterning of them. The facile synthesized property and its natural characteristics make ZnO NRs a widely used template material in the field of sensitized solar cells and preparing nanotubes.

The attractive characteristic of ZnO is the superior electron mobility, which is more than one magnitude order larger than that of the anatase titanium oxide, in all of the semiconductors with the wide bandgap which are taken as replacements of titanium oxide for the electron conductor. ZnO NR-based solar cells are promising devices for solar energy conversion. Because NRs have strong light absorption and rapid carrier collection, in addition they are inexpensive due to the cheap element and small amount of material needed. Compared to planar solar cells, NR photovoltaic devices have enhanced optical absorption due to three effects. ZnO NRs can reduce reflectivity, and the incoming light is captured and confined into guided modes which lead to concentration of the electromagnetic field inside the absorbing material. Moreover, the nanowire arrays support the light along a diffusive path leading to multiple scattering between the wires.

5.2 Excitonic solar cells

Conventional solar cells are the silicon *p-n* junction type invented in the 1950s. Nevertheless, the cost of solar power is too high to be extended industrially. To reduce the cost, a great deal of research has been devoted to less expensive types of solar cell. One of the great promises is the emergence of excitonic solar cells. The difference between conventional and excitonic solar cells is that light absorption results in the formation of excitons in semiconductor materials rather than free electron-hole pairs. Excitonic solar cells consist of molecular semiconductor solar cells, conducting polymer solar cells, dye-sensitized solar cells (DSSCs) [77, 78], and quantum dot solar cells (QDSSCs) [79–82].

Among the different types of excitonic solar cells, the ZnO NR array is popular in the fields of DSSCs and QDSSCs. The dye-sensitized solar cell concept is on the basis of the dye optical excitation, and the conduction band with the metal oxide in the nanostructure wide bandgap is injected into an electron. At the beginning, a kind of the dye-sensitized cell made of a dense array of oriented, crystalline ZnO NWs was researched and attained with a full sun efficiency of 1.5%. To increase the efficiency of such cells, researchers have adopted different methods such as using alternative sensitizers and redox electrolytes to fabricate solid-state or nonvolatile-liquid DSSCs. The new record power conversion efficiency (PCE) in DSCs is 7%, adopted with the synthesized multilayer assemblies of high-surface-area ZnO NWs to fabricate DSSCs [83, 84].

However, despite the successes of DSSCs, novel hybrids of the architectures of device and materials are still hunting to further enhance solar cell performance and cost. Quantum dots (QDs) are one possibility to substitute photosensitive dyes.

Compared to dye, the particle size of QDs can be tuned for adjusting their absorption spectrum to match the solar spectrum better. Also, the efficiency of the photovoltaic (PV) device can be improved effectively by QDs which can make multiple electron-hole pairs per photon. The maximum thermodynamic conversion efficiency of QDSSCs can theoretically reach 44% which is much higher than for DSSCs. In 2007, Aydil ES's group demonstrated ZnO NWs with CdSe QDs photosensitization and provided proof of QDs photogenerating electron transfer to the nanowires for the first time. They proved the possibility of QDs that demonstrated ZnO NWs providing a promising solar cell architecture [85]. Most reported values of ZnO NWs' QDSSCs (typically below 3%) are well below DSSCs (7%). With the time flying, more research on ZnO NWs' QDSSCs keeps forward with the higher efficiency. The performances of QDSSCs are typically limited by problems of aggregation, low QD loading density, and high expense of synthesis to hinder its large-scale applications [86–88].

6. Conclusion

ZnO being one-dimensional (1D) nanostructures is playing an increasingly crucial role in the developing nanoscience and nanotechnology. Due to its unique physical and chemical properties, 1D ZnO nanostructures can definitely enhance remarkably the efficacy in optical sensitivity, (photo)catalysis, mechanical strength, and (thermal and electrical) conductivity, which is beneficial to electronic and energy storage devices, sensors, advanced mechanical materials, and catalysts. Nowadays, the state-of-art resources of the renewable alternative energy with the cutting-edge techniques are urgent to be explored to make them to play the crucial role in the energy consumptions for the future along with the eco-friendly to benefits of the environment and technics, especially considering on the basis of taking an ideal candidate for the traditional energy resources. Solar energy is the radiant energy that is produced by the sun. In many parts of the world, the direct solar radiation is considered to be one of the best prospective sources of energy with the highlighted environmentally friendly benefits. Therefore, the deep insight into the properties of 1D ZnO nanostructures shall be explored more and coupled with the relevant techniques of solar cells. Meanwhile, the methodologies for achieving 1D ZnO nanostructures eco-friendly or green to the environments shall be also researched further along with the relevant detailed mechanism revelation.

Author details

Yeeli Kelvii Kwok
Department of Mechanical and Biomedical Engineering, City University of Hong Kong, Kowloon Tong, Kowloon, Hong Kong

*Address all correspondence to: yeelikwok@yahoo.com

References

[1] Rawtani D, Khatri N, Tyagi S, Pandey G. Nanotechnology-based recent approaches for sensing and remediation of pesticides. Journal of Environmental Management. 2018;**206**:749-762

[2] Kesharwani P, Gorain B, Low SY, Tan SA, Ling ECS, Lim YK, et al. Nanotechnology based approaches for anti-diabetic drugs delivery. Diabetes Research and Clinical Practice. 2018;**136**:52-77

[3] Contreras JE, Rodriguez EA, Taha-Tijerina J. Nanotechnology applications for electrical transformers—A review. Electric Power Systems Research. 2017;**143**:573-584

[4] Kumar MS, Das AP. Emerging nanotechnology based strategies for diagnosis and therapeutics of urinary tract infections: A review. Advances in Colloid and Interface Science. 2017;**249**:53-65

[5] Iavicoli I, Leso V, Beezhold DH, Shvedova AA. Nanotechnology in agriculture: Opportunities, toxicological implications, and occupational risks. Toxicology and Applied Pharmacology. 2017;**329**:96-111

[6] Hussein AK. Applications of nanotechnology to improve the performance of solar collectors-recent advances and overview. Renewable and Sustainable Energy Reviews. 2016;**62**:767-792

[7] Kasaeian A, Eshghi AT, Sameti M. A review on the applications of nanofluids in solar energy systems. Renewable and Sustainable Energy Reviews. 2015;**43**:584-598

[8] Mihindukulasuriya SDF, Lim LT. Nanotechnology development in food packaging: A review. Trends in Food Science & Technology. 2014;**40**(2):149-167

[9] Gupta MC, Ungaro C, Foley JJ, Gray SK. Optical nanostructures design, fabrication, and applications for solar/thermal energy conversion. Solar Energy. 2018;**165**:100-114

[10] Jagadale SB, Patil VL, Vanalakar SA, Patil PS, Deshmukh HP. Preparation, characterization of 1D ZnO nanorods and their gas sensing properties. Ceramics International. 2018;**44**(3):3333-3340

[11] Leitner J, Bartůněk V, Sedmidubský D, Jankovský O. Thermodynamic properties of nanostructured ZnO. Applied Materials Today. 2018;**10**:1-11

[12] Chongsri K, Boonyarattanakalin K, Pecharapa W. Effects of seeding layer thickness on physical and morphological structure of Ga/F co-doped ZnO nanostructures. Taterials Today: Proceedings. 2017;**4**(5, Part 2):6129-6133

[13] Qi B, Ólafsson S, Gíslason HP. Vacancy defect-induced d0 ferromagnetism in undoped ZnO nanostructures: Controversial origin and challenges. Progress in Materials Science. 2017;**90**:45-74

[14] Buller S, Strunk J. Nanostructure in energy conversion. Journal of Energy Chemistry. 2016;**25**(2):171-190

[15] Tereshchenko A, Bechelany M, Viter R, Khranovskyy V, Smyntyna V, Starodub N, et al. Optical biosensors based on ZnO nanostructures: Advantages and perspectives. A review. Sensors and Actuators B: Chemical. 2016;**229**:664-677

[16] Maçaira J, Andrade L, Mendes A. Review on nanostructured photoelectrodes for next generation dye-sensitized solar cells. Renewable and Sustainable Energy Reviews. 2013;**27**:334-349

[17] Spencer MJS. Gas sensing applications of 1D-nanostructured zinc oxide: Insights from density functional theory calculations. Progress in Materials Science. 2012;**57**(3):437-486

[18] Ramos-Ramón JA, Pal U, Cremades A, Maestre D. Effect of Ga incorporation on morphology and defect structures evolution in VLS grown 1D In_2O_3 nanostructures. Applied Surface Science. 2018;**439**:1010-1018

[19] Huang PS, Gao TC. Current development of 1D and 2D metallic nanomaterials for the application of transparent conductors in solar cells: Fabrication and modeling. Nano-Structures & Nano-Objects. 2018;**15**:119-139

[20] Terasako T, Kohno K, Yagi M. Vapor-liquid-solid growth of SnO_2 nanowires utilizing alternate source supply and their photoluminescence properties. Thin Solid Films. 2017;**644**:3-9

[21] Mijangos C, Hernández R, Martín J. A review on the progress of polymer nanostructures with modulated morphologies and properties, using nanoporous AAO templates. Progress in Polymer Science. 2016;**54-55**:148-182

[22] Kente T, Mhlanga SD. Gallium nitride nanostructures: Synthesis, characterization and applications. Journal of Crystal Growth. 2016;**444**:55-72

[23] Gaikwad MA, Suryawanshi MP, Maldar PS, Dongale TD, Moholkar AV. Nanostructured zinc oxide photoelectrodes by green routes M-SILAR and electrodeposition for dye sensitized solar cell. Optical Materials. 2018;**78**:325-334

[24] Tabrizi AG, Arsalani N, Namazi H, Ahadzadeh I. Vanadium oxide assisted synthesis of polyaniline nanoarrays on graphene oxide sheets and its application in supercapacitors. Journal of Electroanalytical Chemistry. 2017;**798**:34-41

[25] Mirzaei A, Neri G. Microwave-assisted synthesis of metal oxide nanostructures for gas sensing application: A review. Sensors and Actuators B: Chemical. 2016;**237**:749-775

[26] Tsega M, Kuo DH, Dejene FB. Growth and green defect emission of ZnPbO nanorods by a catalyst-assisted thermal evaporation-oxidation method. Journal of Crystal Growth. 2015;**415**:106-110

[27] Mukasyan AS, Rogachev AS, Aruna ST. Combustion synthesis in nanostructured reactive systems. Advanced Powder Technology. 2015;**26**(3):954-976

[28] Bueno C, Maestre D, Díaz T, Juárez H, Pacio M, Cremades A, et al. High-yield growth of Ti doped ZnO nano- and microstructures by a vapor-solid method. Journal of Alloys and Compounds. 2017;**726**:201-208

[29] Nersisyan HH, Lee JH, Ding JR, Kim KS, Manukyan KV, Mukasyan AS. Combustion synthesis of zero-, one-, two- and three-dimensional nanostructures: Current trends and future perspectives. Progress in Energy and Combustion Science. 2017;**63**:79-118

[30] Balela MDL, Acedera RA, Flores CLI, Pelicano CMO. Surface modification of ZnO nanostructured film prepared by hot water oxidation. Surface and Coatings Technology. 2018;**340**:199-209

[31] Alinauskas L, Brooke E, Regoutz A, Katelnikovas A, Raudonis R, Yitzchaik S, et al. Nanostructuring of SnO_2 via solution-based and hard template assisted method. Thin Solid Films. 2017;**626**:38-45

[32] Veluswamy P, Sathiyamoorthy S, Chowdary KH, Muthusamy O, Krishnamoorthy K, Takeuchi T, et al. Morphology dependent thermal conductivity of ZnO nanostructures prepared via a green approach. Journal of Alloys and Compounds. 2017;**695**:888-894

[33] Montazerozohori M, Masoudiasl A, Doert T. Two new 1D zigzag Hg(II) nanostructure coordination polymers: Sonochemical synthesis, thermal study, crystal structure and Hirshfeld surface analysis. Inorganica Chimica Acta. 2016;**443**:207-217

[34] Doganay D, Coskun S, Kaynak C, Unalan HE. Electrical, mechanical and thermal properties of aligned silver nanowire/polylactide nanocomposite films. Composites Part B: Engineering. 2016;**99**:288-296

[35] Ghosh S, Dev BN. Probing of O_2 vacancy defects and correlated magnetic, electrical and photoresponse properties in indium-tin oxide nanostructures by spectroscopic techniques. Applied Surface Science. 2018;**439**:891-899

[36] Paek J, Kim J, An BW, Park J, Ji S, Kim SY, et al. Stretchable electronic devices using graphene and its hybrid nanostructures. FlatChem. 2017;**3**:71-91

[37] Matysiak W, Tański T, Zaborowska M. Manufacturing process, characterization and optical investigation of amorphous 1D zinc oxide nanostructures. Applied Surface Science. 2018;**442**:382-389

[38] Abdeltwab E, Taher FA. Polar and nonpolar self-assembled Co-doped ZnO thin films: Structural and magnetic study. Thin Solid Films. 2017;**636**:200-206

[39] Thiebaud L, Legeai S, Stein N. Tuning the morphology of Te one-dimensional nanostructures by template-free electrochemical deposition in an ionic liquid. Electrochimica Acta. 2016;**197**:300-306

[40] Vadivel S, Naveen AN, Theerthagiri J, Madhavan J, Priya TS, Balasubramanian N. Solvothermal synthesis of $BiPO_4$ nanorods/MWCNT (1D-1D) composite for photocatalyst and supercapacitor applications. Ceramics International. 2016;**42**(12):14196-14205

[41] Pal B, Dhara S, Giri PK, Sarkar D. Evolution of room temperature ferromagnetism with increasing 1D growth in Ni-doped ZnO nanostructures. Journal of Alloys and Compounds. 2015;**647**:558-565

[42] Ghosh R, Das R, Giri PK. Label-free glucose detection over a wide dynamic range by mesoporous Si nanowires based on anomalous photoluminescence enhancement. Sensors and Actuators B: Chemical. 2018;**260**:693-704

[43] Chandola S, Speiser E, Esser N, Appelfeller S, Franz M, Dähne M. Optical anisotropy of quasi-1D rare-earth silicide nanostructures on Si(001). Applied Surface Science. 2017;**399**:648-653

[44] Maryasin V, Bucci D, Rafhay Q, Panicco F, Michallon J, Kaminski-Cachopo A. Technological guidelines for the design of tandem III-V nanowire on Si solar cells from opto-electrical simulations. Solar Energy Materials and Solar Cells. 2017;**172**:314-323

[45] Legesse M, Fagas G, Nolan M. Modifying the band gap and optical properties of germanium nanowires by surface termination. Applied Surface Science. 2017;**396**:1155-1163

[46] Vishnukumar P, Vivekanandhan S, Misra M, Mohanty AK. Recent advances and emerging opportunities in phytochemical synthesis of ZnO nanostructures. Materials Science

in Semiconductor Processing. 2018;**80**:143-161

[47] Vittal R, Ho KC. Zinc oxide based dye-sensitized solar cells: A review. Renewable and Sustainable Energy Reviews. 2017;**70**:920-935

[48] Ghoderao KP, Jamble SN, Kale RB. Influence of pH on hydrothermally derived ZnO nanostructures. Optik. 2018;**156**:758-771

[49] Huang MH, Wu YY, Feick H, Tran N, Weber E, Yang PD. Catalytic growth of zinc oxide nanowires by vapor transport. Advanced Materials. 2001;**13**(2):113-116

[50] Huang MH, Mao S, Feick H, Yan HQ, Wu YY, Kind H, et al. Room-temperature ultraviolet nanowire nanolasers. Science. 2001;**292**:1897-1899

[51] Alsultany FH, Hassan Z, Ahmed NM, Almessiere MA. Catalytic growth of one-dimensional single-crystalline ZnO nanostructures on glass substrate by vapor transport. Ceramics International. 2017;**43**:610-616

[52] Athma PV, Martinez AI, Johns N, Safeera TA, Reshmi R, Anila EI. Nanostructured zinc oxide thin film by simple vapor transport deposition. Superlattices and Microstructures. 2015;**85**:379-384

[53] Hosseini ZS, Mortezaali A, Iraji Zad A. Comparative study of the grown ZnO nanostructures on quartz and alumina substrates by vapor phase transport method without catalyst: Synthesis and acetone sensing properties. Sensors and Actuators A. 2014;**212**:80-86

[54] Babu ES, Kim SJ, Song JH, Hong SK. Effects of growth pressure on morphology of ZnO nanostructures by chemical vapor transport. Chemical Physics Letters. 2016;**658**:182-187

[55] Trad TM, Donley KB, Look DC, Eyink KG, Tomich DH, Taylor CR. Low temperature deposition of zinc oxide nanoparticles via zinc-rich vapor phase transport and condensation. Journal of Crystal Growth. 2010;**312**:3675-3679

[56] Haldar SR, Nayak A, Chini TK, Bhunia S. Strong temperature and substrate effect on ZnO nanorod flower structures in modified chemical vapor condensation growth. Current Applied Physics. 2010;**10**:942-946

[57] Pan ZW, Dai ZR, Wang ZL. Nanobelts of semiconducting oxides. Science. 2001;**291**:1947-1949

[58] Tonezzer M, Dang TTL, Bazzanella N, Nguyen VH, Iannotta S. Comparative gas-sensing performance of 1D and 2D ZnO nanostructures. Sensors and Actuators B: Chemical. 2015;**220**:1152-1160

[59] Manthina V, Agrios AG. Single-pot ZnO nanostructure synthesis by chemical bath deposition and their applications. Nano-Structures & Nano-Objects. 2016;**7**:1-11

[60] Bakhsheshi-Rad HR, Hamzah E, Ismail AF, Aziz M, Kasiri-Asgarani M, Akbari E, et al. Synthesis of a novel nanostructured zinc oxide/baghdadite coating on Mg alloy for biomedical application: *In-vitro* degradation behavior and antibacterial activities. Ceramics International. 2017;**43**:14842-14850

[61] Shirahata T, Kawaharamura T, Fujita S, Orita H. Transparent conductive zinc-oxide-based films grown at low temperature by mist chemical vapor deposition. Thin Solid Films. 2015;**597**:30-38

[62] Laurenti M, Garino N, Porro S, Fontana M, Gerbaldi C. Zinc oxide nanostructures by chemical vapour deposition as anodes for Li-ion batteries. Journal of Alloys and Compounds. 2015;**640**:321-326

[63] Menzel A, Subannajui K, Bakhda R, Wang YB, Thomann R, Zacharias M. Tuning the growth mechanism of ZnO nanowires by controlled carrier and reaction gas modulation in thermal CVD. The Journal of Physical Chemistry Letters. 2012;**3**:2815-2821

[64] Park WI, Kim DH, Jung SW, Yi GC. Metalorganic vapor-phase epitaxial growth of vertically well-aligned ZnO nanorods. Applied Physics Letters. 2002;**80**(22):4232-4234

[65] Wu YF, Liu DP, Yu NS, Liu YD, Liang HW, Du GT. Structure and electrical characteristics of zinc oxide thin films grown on Si (111) by metal-organic chemical vapor deposition. Journal of Materials Science and Technology. 2013;**29**(9):830-834

[66] Kim AY, Jang SS, Lee DH, Yim SY, Byun DJ. Effects of temperature on ZnO hybrids grown by metal-organic chemical vapor deposition. Materials Research Bulletin. 2012;**47**:2888-2890

[67] Torres-Huerta AM, Domínguez-Crespo MA, Brachetti-Sibaja SB, Arenas-Alatorre J, Rodríguez-Pulido A. Effect of the substrate on the properties of ZnO–MgO thin films grown by atmospheric pressure metal-organic chemical vapor deposition. Thin Solid Films. 2011;**519**:6044-6052

[68] Fragalà ME, Aleeva Y, Malandrino G. Effects of metal-organic chemical vapour deposition grown seed layer on the fabrication of well aligned ZnO nanorods by chemical bath deposition. Thin Solid Films. 2011;**519**:7694-7701

[69] Wu CL, Chang L, Chen HG, Lin CW, Chang TF, Chao YC, et al. Growth and characterization of chemical-vapor-deposited zinc oxide nanorods. Thin Solid Films. 2006;**498**:137-141

[70] Vallejos S, Gràcia I, Lednický T, Vojkuvka L, Figueras E, Hubálek J, et al. Highly hydrogen sensitive micromachined sensors based on aerosol-assisted chemical vapor deposited ZnO rods. Sensors and Actuators B. 2018;**268**:15-21

[71] Chandrasekhar KR, Arbuj SS. Solvothermal synthesis of one dimensional ZnO nanostructures and its photocatalytic applications. Materials Today: Proceedings. 2015;**2** (9, Part 1)

[72] Badnore AU, Pandit AB. Effect of pH on sonication assisted synthesis of ZnO nanostructures: Process details. Chemical Engineering and Processing Process Intensification. 2017;**122**:235-244

[73] Colorado SA, Colorado HA. Manufacturing of zinc oxide structures by thermal oxidation processes as scalable methods towards inexpensive electric generators. Ceramics International. 2017;**43**:15846-15855

[74] Hong SH, Kim MH, Yun HW, Paik TJ, Lee H. Solution-processed fabrication of superhydrophobic hierarchical zinc oxide nanostructures *via* nanotransfer printing and hydrothermal growth. Surface & Coatings Technology. 2017;**331**:189-195

[75] Chen RQ, Zou CW, Yan XD, Gao W. Zinc oxide nanostructures and porous films produced by oxidation of zinc precursors in wet-oxygen atmosphere. Progress in Natural Science: Materials International. 2011;**21**:81-96

[76] Kumarakuru H, Urgessa ZN, Olivier EJ, Botha JR, Venter A, Neethling JH. Growth of ZnS-coated ZnO nanorod arrays on (100) silicon substrate by two-step chemical synthesis. Journal of Alloys and Compounds. 2014;**612**:154-162

[77] Sinha D, De D, Goswami D, Ayaz A. Fabrication of DSSC with nanostructured ZnO photo anode and

natural dye sensitizer. Materials Today: Proceedings. 2018;**5**(1, Part 2):2056-2063

[78] Marimuthu T, Anandhan N, Thangamuthu R. Electrochemical synthesis of one-dimensional ZnO nanostructures on ZnO seed layer for DSSC applications. Applied Surface Science. 2018;**428**:385-394

[79] Nayeri FD, Akbarnejad E, Ghoranneviss M, Soleimani EA, Hashemizadeh SA. Dye decorated ZnO-NWs/CdS-NPs heterostructures for efficiency improvement of quantum dots sensitized solar cell. Superlattices and Microstructures. 2016;**91**:244-251

[80] Eskandari M, Ahmadi V, Kohnehpoushi S, Yousefi rad M. Improvement of ZnO nanorod based quantum dot (cadmium sulfide) sensitized solar cell efficiency by aluminum doping. Physica E: Low-dimensional Systems and Nanostructures. 2015;**66**:275-282

[81] Rawal SB, Sung SD, Moon SY, Shin YJ, Lee WI. Optimization of CdS layer on ZnO nanorod arrays for efficient CdS/CdSe co-sensitized solar cell. Materials Letters. 2012;**82**:240-243

[82] Li L, Zhai TY, Bando Y, Golberg D. Recent progress of one-dimensional ZnO nanostructured solar cells. Nano Energy. 2012;**1**(1):91-106

[83] Sharma D, Jha R, Kumar S. Quantum dot sensitized solar cell: Recent advances and future perspectives in photoanode. Solar Energy Materials and Solar Cells. 2016;**155**:294-322

[84] Jafarzadeh M, Sipaut CS, Dayou J, Mansa RF. Recent progresses in solar cells: Insight into hollow micro/nano-structures. Renewable and Sustainable Energy Reviews. 2016;**64**:543-568

[85] Leschkies KS, Divakar R, Basu J, Enache-Pommer E, Boercker JE, Carter CB, et al. Photosensitization of ZnO nanowires with CdSe quantum dots for photovoltaic devices. Nano Letters. 2007;**7**(6):1793-1798

[86] Xu YF, Wu WQ, Rao HS, Chen HY, Kuang DB, Su CY. CdS/CdSe co-sensitized TiO_2 nanowire-coated hollow spheres exceeding 6% photovoltaic performance. Nano Energy. 2015;**11**:621-630

[87] Sun JK, Jiang Y, Zhong XH, Hu JS, Wan LJ. Three-dimensional nanostructured electrodes for efficient quantum-dot-sensitized solar cells. Nano Energy. 2017;**32**:130-156

[88] Majumder T, Dhar S, Chakraborty P, Debnath K, Mondal SP. Advantages of ZnO nanotaper photoanodes in photoelectrochemical cells and graphene quantum dot sensitized solar cell applications. Journal of Electroanalytical Chemistry. 2018;**813**:92-101

Section 3

Applications

Chapter 5

Biodegradable Chitosan Matrix Composite Reinforced with Titanium Dioxide for Biocidal Applications

Johnny López Calero, Zuleika Oquendo Berríos and Oscar M. Suarez

Abstract

Access to drinkable water is increasingly difficult in developing nations. According to the World Health Organization, more than 1.2 billion people lack access to this resource. The scientific community has risen to the challenge by developing innovating methods to purify water. In this context, a chitosan/titanium dioxide biocomposite represents an appealing, economically viable solution to remove oils, heavy metals, and pathogen colonies from contaminated waters. The present chapter reports the study of a chitosan film matrix impregnated with nanoparticles, which was capable of expressing antibacterial properties when exposed to UV light. Several techniques allowed characterizing the biocomposite uniformity, corroborating the crystal structure and assessing the bonding type of this biocomposite mixture. Antibacterial tests with *Escherichia coli* (Gram-negative) and *Staphylococcus aureus* (Gram-positive) were completed via growth curve analysis and the Kirby-Bauer technique. The results of this encompassing study revealed that bacterial growth was reduced by more than 50%. Optical microscopy, Fourier-transform infrared spectroscopy, and X-ray diffraction helped to characterize this biocomposite film. This chapter discusses how a novel and biodegradable film represents a better antibacterial material that is able to eliminate pathogens cost-effectively.

Keywords: chitosan, titanium dioxide, anatase, *Escherichia coli*, *Staphylococcus aureus*, composite, biocide

1. Introduction

The widespread use of antibiotics has promoted the dissemination of multidrug resistance (MDR) genes in bacteria, whereas nanoparticles (NPs) have proven to be effective when dealing with this bacteria type [1]. Superbugs (i.e., bacteria which are resistant to nearly all antibiotics) have evolved due to the abuse, misuse, and negligence of antibiotics. They are capable of transferring their resistance by mechanisms as plasmid [2], transposon [3], and integrons [4]. This means that just one superbug can render an infection nearly untreatable. Antibiotics target mainly three areas inside bacteria: the cell wall, the translational machinery, and the DNA replication system. Superbugs can develop genes capable of resisting the effect of antibiotics, ranging from the expression of enzymes and modification of cell components all the way to the creation of efflux pumps [5]. For example, enzymes such as β-lactamase [6] could alter the chemical structure of the antibiotics, leaving them inactive. Meanwhile, cell components such as ribosomes [7] offer resistance and efflux pump work by removing antibiotics from the inside of a cell and maintaining a safe environment for which the bacteria can grow normally.

Furthermore, NPs work differently when it comes to dealing with bacteria; this makes any antibiotic resistance ineffective against NPs. The NP antimicrobial mechanism falls into three categories: oxidative stress induction [8], metal ion release [9], and nonoxidative [10]. These mechanisms all work toward producing bacterial cell death.

Nanoparticles, as nanostructured materials, can be defined as three-dimensional materials with dimensions in the nanoscale (1–100 nm). As many studies suggest, they offer an array of antibacterial properties against Gram-negative and Gram-positive bacteria, such as *E. coli* and *S. aureus*, respectively. For example, zinc oxide (ZnO) NPs are capable of inhibiting the growth of *S. aureus*, while silver NPs demonstrate antimicrobial activity against *E. coli* [11].

As mentioned above, there are three main mechanisms by which NPs produce apoptosis. One can summarize the entire process as follows: first, the disruption of bacterial cell membrane occurs, followed by (whenever possible) the production of reactive oxygen species (ROS), which induces the penetration of the bacterial cell to disrupt the intracellular mechanism needed by the bacteria to function properly. NPs can produce ROS by different mechanisms, but, in the scientific community, the photocatalytic hypothesis is most widely accepted.

MDR genes have proliferated to such an extent that they become a challenge for the existing resources to counteract them. Accordingly, investigations related to antibacterial materials have become mandatory to protect human lives.

Prior research demonstrated that separately chitosan and TiO_2 (anatase polymorph) possess antibacterial properties under ultraviolet light irradiation. This characteristic can clean water sources, which represents one of the main vectors used by bacteria to infect the human body.

Water has become a precious resource, causing the access to fresh and clean water to become a critical matter at a global scale. Nowadays, innovative methods must offer clean, fresh, and purified water. As mentioned above, in this context, chitosan and TiO_2 biocomposites could be an economically viable alternative to resolve this problem by removing oils, heavy metals, and pathogen colonies.

2. Bacteria

Because of their incidence in global health, this study centered on two types of bacteria: a Gram-negative and a Gram-positive one. Details of them are provided in the next sections.

2.1 *Escherichia coli*

Escherichia coli (*E. coli*) is a Gram-negative, rod-shaped, facultative anaerobic bacterium, as shown in **Figure 1**. The cell wall consists of an outer membrane with lipopolysaccharide and peptidoglycan along with a plasma membrane. A good analogy is that this type of bacteria bears both an outer and an inner skin that offer double protection against foreign agents. Theodor Escherich first described this microorganism in 1885, making it one of the most studied bacteria in the world. *E. coli* strains colonized the gastrointestinal tract of humans and animals as a normal

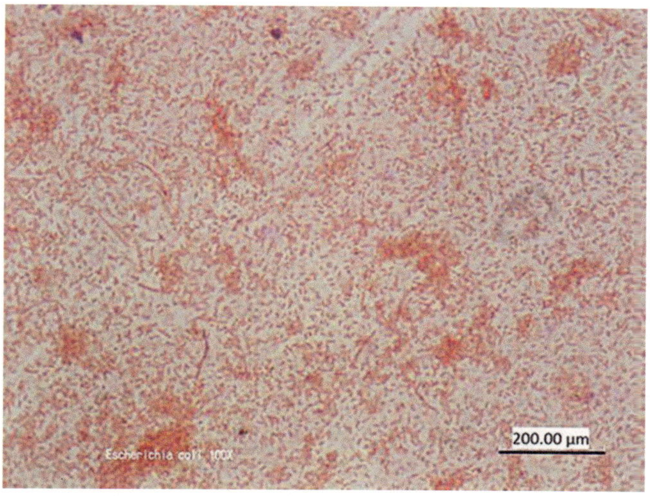

Figure 1.
Optical image of E. coli stained using Gram's method.

flora [12]. Despite this, some strains have evolved into a pathogenic *E. coli*. This bacterium acquires virulence factors through transposons, bacteriophages, plasmids, and other pathogenicity islands. One can categorize *E. coli* as a pathogen, according to the pathogenicity mechanisms and clinical symptom serogroups or through virulence factors [12].

There is a variety of foods and environmental sources for the *E. coli* to grow that are resistant to a breadth of antimicrobial drugs, usually used in medicine and agriculture. Their particular resistance to antibiotics is of great importance because these are Gram-negative pathogens commonly found in humans. For instance, there are different sources of contamination of vegetables, such as poor water treatment and the presence of fertilizers used during cultivation. Similarly, water or food contaminated with wastes can infect animals. Another vector of contamination is via meat products as a result of improper handling and processing or lack of hygiene [13].

There are numerous serotypes in enterohemorrhagic *E. coli* (EHEC) that are frequently associated with human diseases such as O26:H11, O91:H21, O111:H8, O157:NM, and O157:H7 [12]. *E. coli* O157:H7 is the most common isolated serotype of *E. coli*. It has been frequently isolated from people in the United States, Japan, and the United Kingdom. The *E. coli* O157:H7 is also a common food-related pathogen leading to major worldwide health problems. This enterohemorrhagic pathogen can cause different clinical symptoms ranging from diarrhea to death.

2.2 *Staphylococcus aureus*

Staphylococcus aureus (*S. aureus*) is also a common bacterium in the human body. This Gram-positive bacterium possesses a cell wall bearing only a thick peptidoglycan layer followed by a plasma membrane, as shown in **Figure 2**. It is present in the mucous membranes (nose) and in the skin [14]; 30% of healthy adults have these bacteria in the nose, and about 30% has it in the skin [15]. These percentages are increased in people who work in hospitals. *S. aureus* is spread by contact with infected people or by contact with contaminated objects.

Some common infections caused by *S. aureus* are:

- Pneumonia that causes fever, shortness of breath, and cough, causing abscesses in the lungs and accumulation of pus

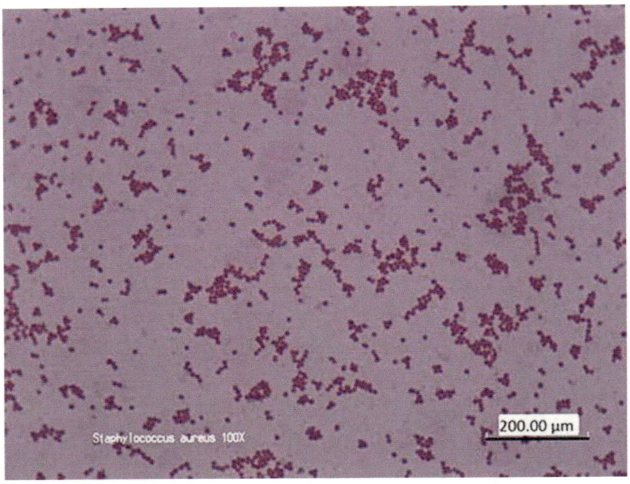

Figure 2.
Optical image of S. aureus stained using Gram's method.

- Toxic epidermal necrolysis: infection that causes detachment of large amounts of skin

- Blood flow infection that produces high and persistent fever and shock

- Osteomyelitis: infection of the bones that can reach the bone through the bloodstream or can spread through the surrounding tissue

UV radiation can inhibit the bacteria growth. This is commonly used in medicine to sterilize surgical instruments. UV radiation kills the bacterium by disrupting its reproduction mechanism [16].

3. Chitin and chitosan overview

In recent years, the use of biopolymers has increased due to their broader application in cosmetics, pharmaceuticals, food industry, biomedical, agriculture, and environmental products. Chitin was discovered in 1811 by Braconnot and chitosan, in 1859, by Rouget [17]. Chitin is made of N-acetyl-D-glucosamine monomer units linked with β(1,4) glycosidic linkage, whereas chitosan is the deacetylated form of chitin [18]. In addition, chitosan presents a rigid crystalline structure due to inter- and intramolecular hydrogen bonds as a result of the amine and hydroxyl groups [18].

Chitosan is a long chain of hydrophilic polysaccharide with a chemical formula of β(1–4) 2-asetamide-2-deoxy-D-glucopyranose. Chitosan is also known to be a natural polymer with similar molecular structure as cellulose; however it varies in the C-2 chain where the hydroxyl group in cellulose is replaced by amine (NH_2). Chitosan is used to inhibit biofouling, being nontoxic and biocompatible, and also possesses inherent antimicrobial properties [19].

Chitin can be extracted from the exoskeletons of crustaceans and fungi by a two-step chemical processing: demineralization and deproteinization [20]. A wide variety of chemical processes allow preparing chitosan: solution casting, dipping and spray coating, compression molding, blending, layer-by-layer, and 3D printing [21].

Chitin along with chitosan is considered the most abundant natural polysaccharides after cellulose. Due to the abundance of waste of shrimps, crabs, and other crustacean exoskeletons, chitin and chitosan are both a cost-effective and abundance product. As polymers, they can be shaped as membranes, films, nanofibers, capsules, tablets, microspheres, pastes, gels, powders, filaments, granules, sponges, and hydrogels. Biomedical uses of chitosan and chitin include drug delivery systems, tissue engineering, and wound healing, as a result of their extensive biocompatibility and biodegradability. For instance, chitosan can be degraded in vertebrates mainly by lysozyme and certain bacterial enzymes in the colon [22]. Moreover, chitosan presents a high in vivo biocompatibility and biodegradability with minimal inflammation and body reaction when applied as tissue [23]. Moreover, the film adhesion on tissue can induce the proliferation of fibroblasts, indicating high biocompatibility [23]. Despite the great biological properties, such as biocompatibility, nonimmunogenicity, antibacterial, and anti-tumorigenicity, there is a lack of sufficient knowledge on the regenerative mechanism of this biopolymer [24]. Another advantage is their functionalization, especially in the hydrophilic group, hydrophobic group, cationic group, targeting ligand, thiol group, and amino acid. This can lead to various gene therapy applications by improving their solubility, toxicity, buffering capacity, escapes in endosome, cellular uptake, genetic material release, transfection efficiency, and silencing efficiency [17]. Chitosan is also known for its anti-inflammatory effects on in vivo tissue applications, which can be applied to inhibit growths of Gram-positive and Gram-negative bacteria. Acting alone, chitosan has shown much promise in wastewater treatment, as it can be used in the removal of dyes, odor, organic pollutants, and inorganic heavy metal ions from industrial wastewater [21]. It can also be applied as an antifouling material used on boat paint because of its prevention on bacteria formation [25]. A problem is that parameters such as molecular weight of chitosan, acetic acid concentration, sterilization process, thickness of the matrix, pH of the medium, and cell parameters (i.e., age, concentration, and bacteria type) led to important differences on microbial growth [18].

4. Titanium dioxide

Titanium dioxide has many commercially available applications in the food industry as a food additive, in the cosmetic industry as a makeup additive and sunscreen, and in the health industry as a drug delivery system [26]. It has also been used as a treatment of wastewater to remove pollutants [27]. Pure titanium dioxide when exposed to ultraviolet light (wavelength less than 400 nm) becomes a photocatalytic material [28]. This is largely dependent on particle size, shape, surface characteristics, and the dioxide crystal structure [29]. Different TiO_2 crystal structures (viz., anatase, rutile, and brookite) have different photocatalytic properties [29]. Sayes et al. demonstrated that the anatase polymorph is more reactive under ultraviolet light (UV) [30], which renders it more cytotoxic. Furthermore, Blinova et al. stated that pure anatase induces apoptosis but does not generate ROS, while rutile does initiate cell death through the formation of ROS [31] Davis et al. proposed that anatase particles under 100 nm have a strong UV absorption [32].

Nevertheless, TiO_2 is a proven photocatalytic material under UV light. When this energy is greater than the band gap, i.e., 3.06 eV for rutile and 3.23 eV for anatase [33], the electrons in the valence band (VB) are excited to the conductive band (CB), producing a corresponding hole in the valence band and highly reactive reactants (e^- and h^+) on the TiO_2 surface, as shown in **Figure 3**.

The electrons and holes react with water and air to form the highly chemical active ROS. There are four main and common ROS: superoxide radical (O^{2-}), hydroxyl radical (OH^-), hydrogen peroxide (H_2O_2), and singlet oxygen (O^-) [34]. They range from less acute (O^{2-} and H_2O_2) to more acute (O^{2-} and O^-) [34]. Not all

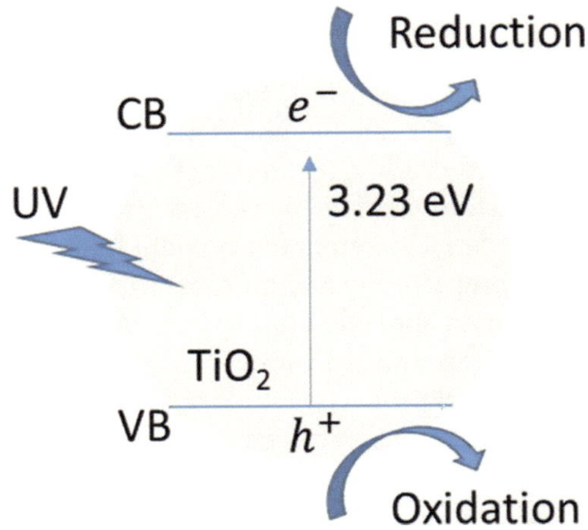

Figure 3.
Schematic illustration for the photocatalytic reaction of TiO_2.

NPs can produce ROS, and even if they could, many do not produce them all. For example, calcium oxide (CaO) and magnesium oxide (MgO) NPs can only produce O^{2-} [34]. ZnO can generate H_2O_2 and OH^- but not O^{2-} [34], while TiO_2 and copper oxide (CuO) NPs can produce all four types of ROS [34]. For the sake of simplicity, the present work focuses only on TiO_2 irradiated with UV light and is used as a ROS source.

Such ROS production induced by the irradiation of TiO_2 with UV light is shown in the following equations. Eq. (1) describes the energy absorption and the photocatalytic reaction. Eqs. (2)–(4) depict the photocatalytic redox pathways involved in the generation of an O^{2-} and an OH^- at the reaction between the holes with H_2O and the electrons with O^{2-} [35]. Eqs. (5), (6) describe the generation of H_2O_2 by reductive and oxidative pathways, respectively.

$$TiO_2 + energy \rightarrow e^-_{CB} + h^+_{VB} \tag{1}$$

$$O_2 + e^-_{CB} \rightarrow O^{2-} \tag{2}$$

$$H_2O + h^+_{VB} \rightarrow OH^- + H^+ \tag{3}$$

$$OH^- + h^+_{VB} \rightarrow OH^- \tag{4}$$

$$O_2^- + 2H^+ + e^-_{CB} \rightarrow H_2O_2 \tag{5}$$

$$2h^+_{VB} + 2H_2O \rightarrow H_2O_2 + 2H^+ \tag{6}$$

4.1 Oxidative stress

The ROS formation is an oxidative stress mechanism due to the generation of an imbalance between the production of free radicals and the ability of the cell to counteract. Different bacteria have particular ways to protect themselves to some degree against ROS. However, any overproduction of this species initiates a systematic failure that starts by damaging the cell component and ends with the cell death. Human cells are not exempt from the effect of ROS [36], due to their self-production in the mitochondria.

In bacteria the reactive oxygen species attack the cell membrane and proteins, affecting the interaction between deoxyribonucleic acid (DNA) and cell and increasing the gene expression for oxidative proteins [37]. Several studies found that the expression of two oxidative stress genes (catalase Kat A and alkyl hydroperoxide reductase, Ahp C) and a general stress response gene (chaperon protein, DNA K) rose by 52, 7, and 17 times, respectively, as revealed by real-time polymerase chain reaction (RT-PCR) [38].

Negatively charged O^{2-} and OH^{-} can exist on the cell surface and cannot penetrate the intracellular regions of the bacteria because of the negative charge barrier; conversely, hydrogen peroxide [39] can easily pass through. Therefore, NPs that produce all ROS have a higher opportunity as a broad-spectrum biocide. Furthermore, there exists evidence that TiO_2 increases the membrane permeability of a bacterial cell, which allows ROS to enter and start the cell death process [40].

5. Methodology

5.1 Materials

For the production of the biocomposite, the following materials were used: chitosan powder (coarse ground flakes, deacetylated chitin, poly(D-glucosamine) CAS number: 9012-76-4), acetic acid (glacial ACS reagent (2.5 L, RABA0010–2.5D1)), and TiO_2 (anatase polymorph) purchased from Sigma Aldrich® now Millipore Sigma® (22 nm, nanocrystalline colloidal paste for transparent films, >95% anatase by x-ray diffraction, 60–65% porosity, specific surface area 65–75 m^2, pH <1, #798495).

To test the antibacterial properties of this renewable biocomposite, *E. coli* (ACTC 25922) and *S. aureus* (ATCC25923), provided by the Biology Department at the University of Puerto Rico—Mayagüez (UPRM), were used. The bacterial medium used to grow them was the Miller's Luria Broth (LB, CAS number: 91079–40-2, tryptone 10 g/L, yeasts 5 g/L, and sodium chloride 10 g/L), provided by Research Product International (RPI). An in-house Millipore filter provided the necessary deionized water. To provide aseptic conditions and for disinfection, common Lysol® diluted in distilled water was used. The use of protective gear, i.e., face masks, laboratory coats, gloves, and safety glasses, was mandatory throughout the entire experimental work. Laboratories in the UPRM Biology Building and Stéfani Building hosted the present research.

5.2 Chitosan/TiO_2 film production

Following the procedure for polymer solution casting, we measured and mixed chitosan powder with acetic acid and deionized water following this schedule: a mechanical agitator for 1.5 h, a sonicator for 1 h, and a magnetic agitator for another 1 h. The TiO_2 powder (anatase polymorph) was added in increments of 0.5 wt% to reach each target concentration and to prepare four different mixtures of TiO_2 and chitosan. These mixtures were left inside an oven at 55°C for 48 hrs to be then utilized and characterized. **Figure 4** shows the as-produced films ready to be tested.

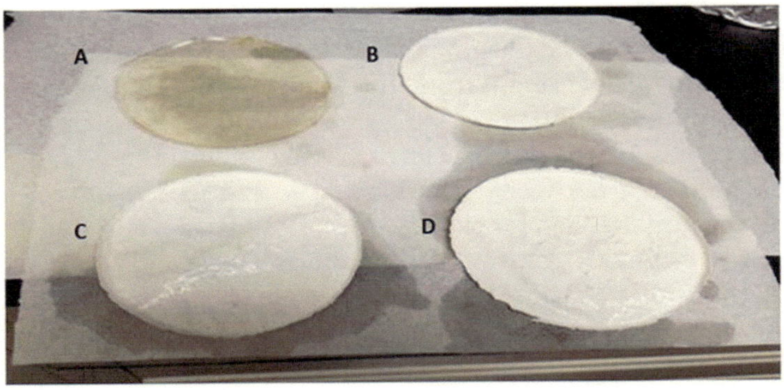

Figure 4.
Films as extracted from the oven. (A) 0.0% TiO_2, (B) 0.5% TiO_2, (C) 1.0% TiO_2, and (D) 1.5% TiO_2.

5.3 Bacteria medium preparation for growth curve analysis

In this protocol, we prepared the bacterial medium of Luria broth (LB) nearly 24 h before each experiment. All laboratory glassware was autoclaved for 15 minutes before use. One liter of water was mixed with 10 g of LB assisted by a magnetic agitator. After obtaining a visually homogenous mixture, we divided it in three groups of laboratory plastic tubes, namely, control, *E. coli*, and *S. aureus*, and autoclaved a second time for 15 minutes. Each bacteria medium was left overnight inside a fume hood under UV light. Two master solutions were separated from the group, incubated with the bacteria, and left inside a shaker at 200 rpm and 37.0°C for almost 6 h. We followed this entire procedure twice per experiment to produce enough samples to be tested with and without UV.

6. Characterization

6.1 Growth curve analysis

Growth curve analysis (GCA) allowed assessing the effect of the ROS generated from the TiO_2 upon UV irradiation. As mentioned, we tested two bacteria: *E. coli* and *S. aureus*. We removed the bacterial media from the hood (protocol described previously) and mixed them with the films impregnated with TiO_2 in laboratory plastic tubes as shown in **Table 1**.

A spectrophotometer allowed measuring the optical density (OD) of each master solution for each bacterium so that one could extract a fixed amount of volume (aliquot) from the solution for each bacterium. This aliquot was then added and mixed to each laboratory plastic tube with the bacteria medium and biocomposite film, as mentioned above. To know exactly how much volume was needed, Eq. [7] was used. Eq. [7] related the volume needed (v_2) of aliquot to add to each plastic tube. C_1, C_2, and V_1 are the concentration of the plastic tube, the concentration in the master solution, and volume of medium already in the plastic tube (20 mL), respectively. The concentration was obtained from the OD measurement.

$$V_2 = \frac{C_1 V_1}{C_2} \quad (7)$$

Table 1 presents the three groups formed, i.e., the control group with only the composite films and the medium without bacteria, the group with *E. coli* and

	Number of laboratory tubes				
Group	*S. aureus*	*E. coli*	Control	TiO_2	UV
Day 1	3	3	3	Yes	Yes
	3	3	3	No	Yes
Day 2	3	3	3	No	No
	3	3	3	Yes	No

Table 1.
The number of laboratory plastic tubes used and how the groups were organized.

films with bacteria medium, and the group with *S. aureus* and films with bacteria medium.

After the three groups were ready, we measured the OD after the first 20 minutes three times (20, 40, and 60 minutes) of the mix preparation, then after 40 minutes three times (100, 140, and 180 minutes), and, finally, after an hour five times (240, 300, 360, 420, and 480 minutes).

During the wait time, i.e., between measurements, the test tubes remained inside the shaker under optimal conditions for bacterial growth. It is important to understand that we completed this entire process twice. As UV light irradiated the samples, we covered the shaker with a black box to block any other light source. **Figures 5** and **6** show how the films without any UV irradiation affected the bacteria growth.

Figures 5 and **6** prove the negligible antibacterial capacity of the films without photoactivation (no UV light). As we can observe in both **Figures 5** and **6**, the samples with bacteria grow normally (if we compared the sample with only bacteria with the samples of bacteria and films), while the samples without bacteria stay constant (0.0 absorbance) until close to the end of the experiment. The samples are either grouped on the top (samples with bacteria) or in the bottom (samples without bacteria). On the other hand, **Figures 7** and **8** (obtained for *S. aureus* and *E. coli*, respectively) have shown that the effect of photoactivated films becomes apparent. The growth rate reduction is attributed to the effects of the ROS liberated from TiO_2 when exposed to energy from the UV light.

Figures 7 and **8** display a difference in the bacteria growth on the films photoactivated by UV. Further, **Figures 9** and **10** (obtained for *S. aureus* and *E. coli*, respectively) evince the difference in growth percent of the bacteria exposed to the

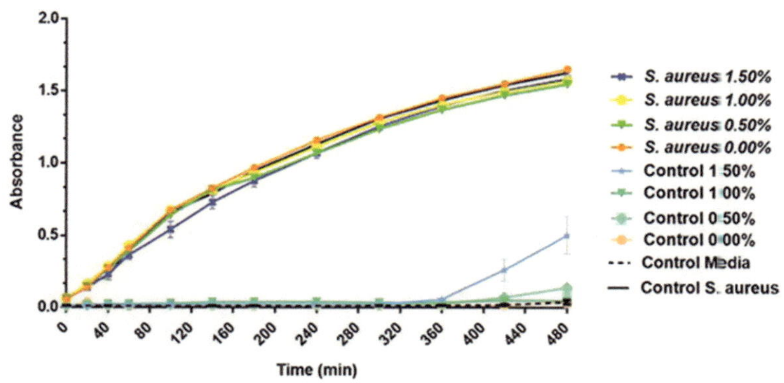

Figure 5.
S. aureus exposed for 8 h to the chitosan films impregnated with TiO_2, without UV irradiation. The upper curves indicate the presence of bacteria, while the bottom ones indicate lower bacteria content.

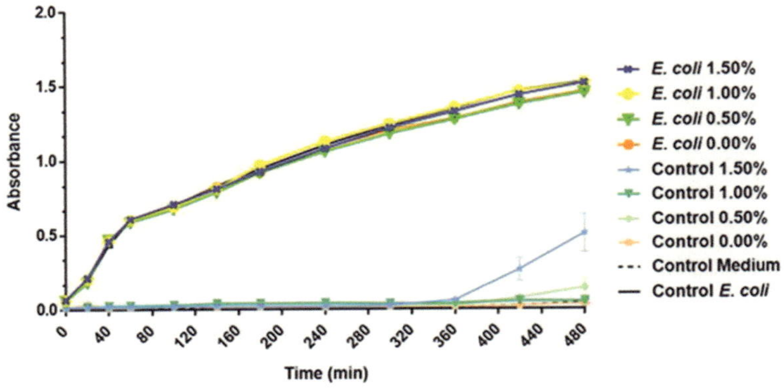

Figure 6.
E. coli exposed for 8 h to the chitosan films impregnated with TiO_2, without UV irradiation. The upper curves indicate the presence of bacteria, while the bottom ones indicate lower bacteria content.

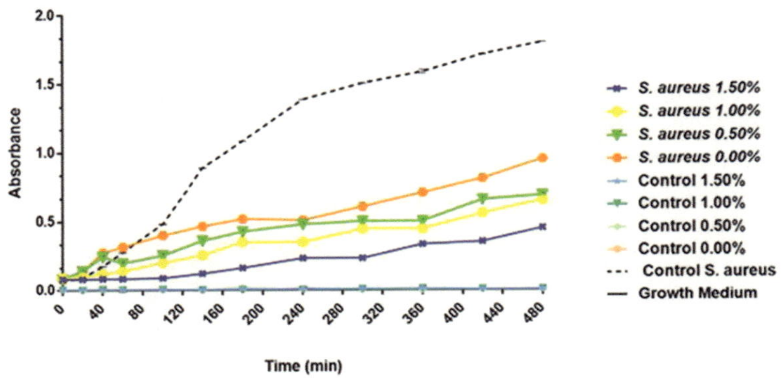

Figure 7.
S. aureus exposed for 8 h to the chitosan films impregnated with TiO_2, with UV irradiation.

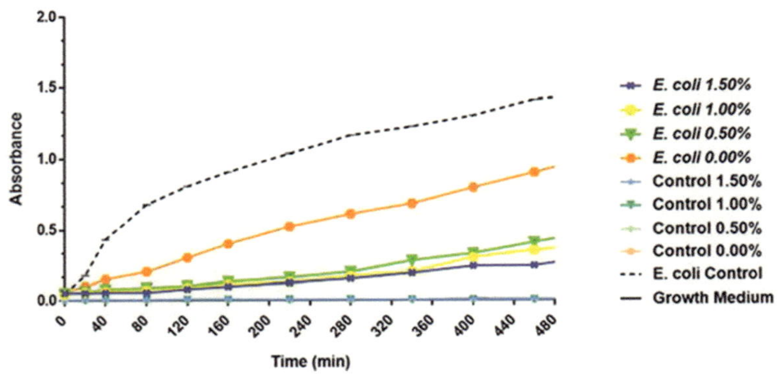

Figure 8.
E. coli exposed for 8 h to the chitosan films impregnated with TiO_2, with UV irradiation.

films after the growth rate stabilized at the end of the experiment. As observed in these two figures, the celling line corresponds to the normal growth of the bacteria. The bars represent the growth rate of each bacterium exposed to the different concentrations of TiO_2 in the films. **Figures 11** and **12** show the same bacteria groups exposed to the same level of TiO_2 but this time with UV light.

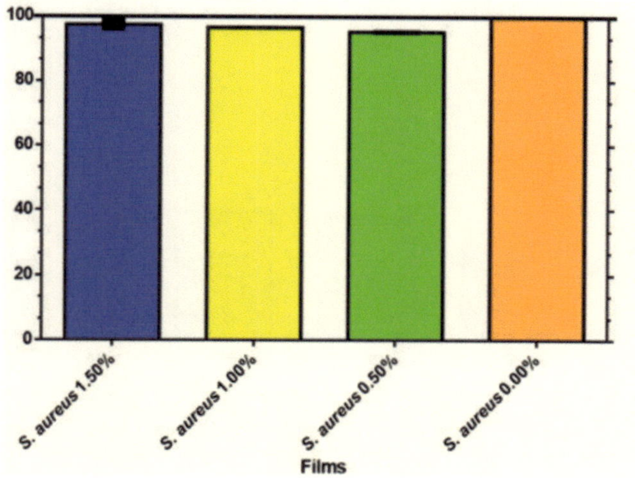

Figure 9.
Growth percent of S. aureus exposed for 8 h to the biocomposite films made of chitosan and TiO_2.

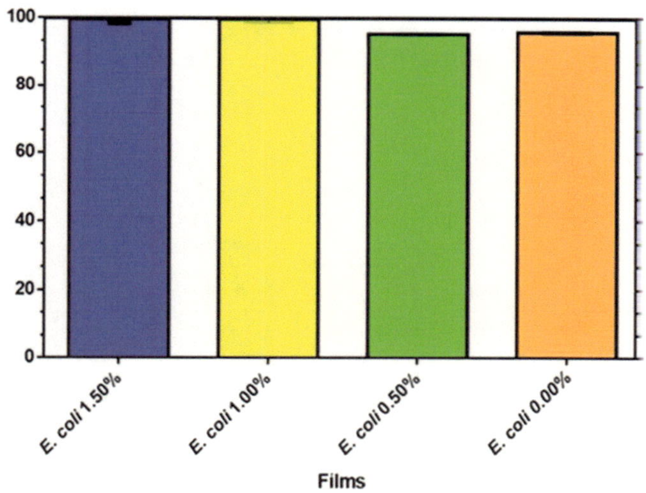

Figure 10.
Growth percent of E. coli exposed for 8 h to the biocomposite films made of chitosan and TiO_2.

As observed in **Figures 9** and **10**, the celling line is the normal growth of the bacteria. The bars are the bacteria exposed to the different concentrations of TiO_2 in the films. **Figures 11** and **12** (obtained for *S. aureus* and *E. coli*, respectively) show the same bacteria groups exposed to the same concentration of TiO_2 but this time with UV light.

Figures 11 and **12** demonstrate how the exposure of the films to the UV light did affect the bacteria growth. One can say that the UV light alone is detrimental to the bacteria but the experiments with only UV exposure did not affect the growth rate significantly.

6.2 Inhibition ring analysis

This analysis, also known as the Kirby-Bauer (KB) method, helps determine the susceptibility to solutions or drugs of isolated microorganisms, i.e., in our case *E. coli* and *S. aureus*. We cut the films as circular pieces using a paper perforator so

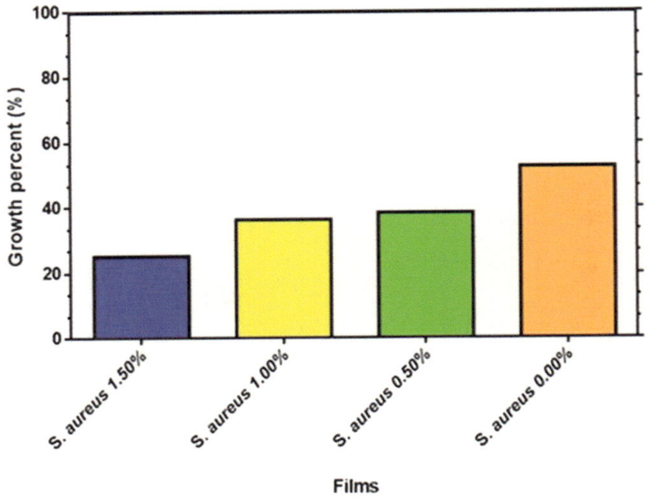

Figure 11.
Growth percentage of S. aureus exposed for 8 h to the biocomposite films made of chitosan and TiO_2 with UV light.

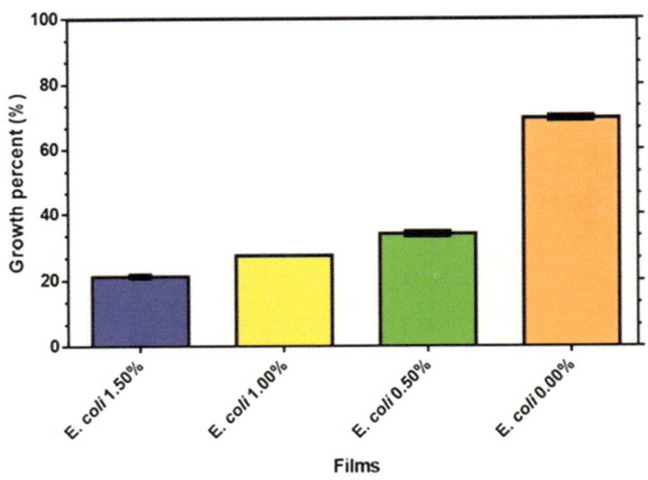

Figure 12.
Growth percentage of E. coli exposed for 8 h to the biocomposite films made of chitosan and TiO_2 with UV light.

that all pieces had the same diameter. These were then exposed to UV irradiation for different times. The best results were obtained on films exposed to UV light for 168 h, as shown in **Figures 13** and **14**. When the UV exposure was over, we placed the films over the bacteria in the Petri dishes. Afterward, the bacteria would grow forming a ring along the chitosan piece with NPs. We measured the area of this ring as a measure of the biocidal potential of the biocomposite. In other words, a larger ring area represented better antibacterial properties of the material.

Figure 13 shows four groups of *S. aureus*. Each group was exposed to a different film with different TiO_2 concentrations. The horizontal axis corresponds to the time elapsed after the films were placed over the bacteria. Naturally, at 0.00 h there was no inhibition ring yet since the films had just been placed on the bacteria.

Similar to **Figure 13**, **Figure 14** has the bacteria *E. coli* exposed to different concentrations of TiO_2. The very first point after the films were placed was without inhibition rings yet. Here the optimal TiO_2 concentration was 1.0% and not 1.5%, which was to be expected due to the double membrane that the *E. coli* possesses.

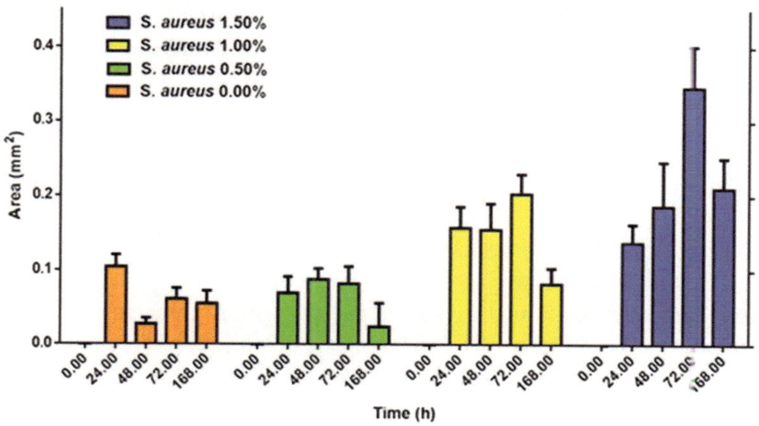

Figure 13.
Area of inhibition ring for S. aureus after films are exposed for 168 h to UV light. Each bar was a different data point of the same plate taken continuously every 24 h.

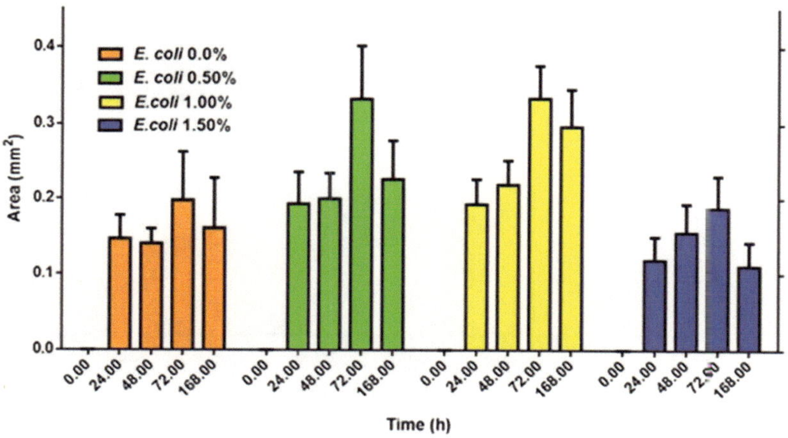

Figure 14.
Area of inhibition ring of E. coli after films are exposed for 168 h to UV light. Each bar was a different data point of the same plate taken continuously every 24 h.

6.3 Fourier-transform infrared (FTIR)

The Fourier-transform infrared (FTIR) spectroscopy helps identify functional groups in a molecule. In our case, we used it to identify the chemical bonds present and how they degraded after UV exposure. As it is well-known, UV irradiation degrades the covalent bonds in polymers, rendering them brittle after a prolonged time under UV light.

In that context, FTIR spectroscopy, as a chemical analytical technique, helps measure the infrared intensity as a function of the wavelength (or wave number) of light. Based upon said wave number, infrared light can be categorized as far infrared, mid infrared, and near infrared for use in polymer science, organic synthesis, petrochemical engineering, food analysis, and even pharmaceutical industries [41, 42]. **Figures 15–18** show the results from four films after the UV exposure. Even more, by using the transmittance percent, we can detect the difference among films before and after such UV exposure.

The characteristic bands of 3400, 3000, 1080, and 1030 cm^{-1} were assigned to the stretching vibration of O—H groups, C—H group, C—N groups, and C—O—C groups, respectively. The band at 1618 cm^{-1} was attributed to the bending vibration of N-H.

A transmittance increment becomes evident after the UV light exposure on the films according to our results in **Figures 15–18**. The breakdown of the covalent bonds in the polymer reveals the degradation. Each valley is a different bond that when degraded absorbs less energy.

6.4 X-ray diffraction (XRD)

X-ray diffraction (XRD) is a phase identification tool in crystal materials [43]. In the present study, XRD analysis allowed confirming the crystalline nature of the TiO_2 used, identifying it as the anatase polymorph. A comparison between two diffraction patterns is in **Figure 19**. Further, this XRD analysis revealed that during the entire synthesis, process of the films (described in Section 5.2) did not affect the original anatase polymorph of the TiO_2. In effect, via **Figure 19**, we can conclude this because each peak (reflected by specific indexed crystallographic planes) aligns

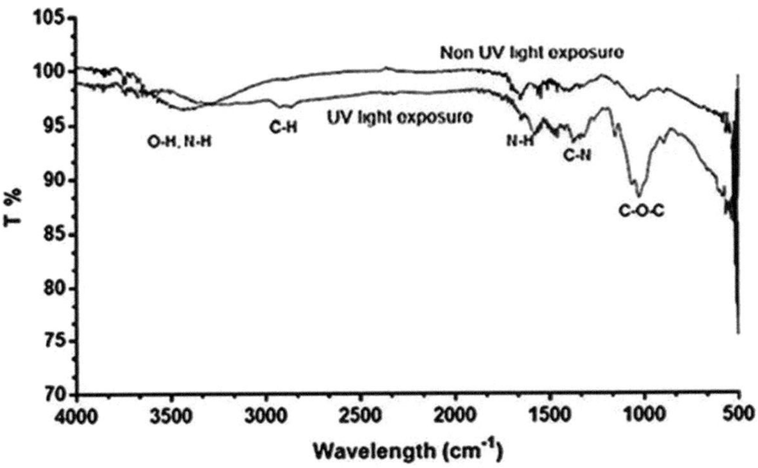

Figure 15.
FTIR spectrum of the 0.0% TiO_2 film after the UV exposure (blue line) for 24 h and without UV exposure (black line).

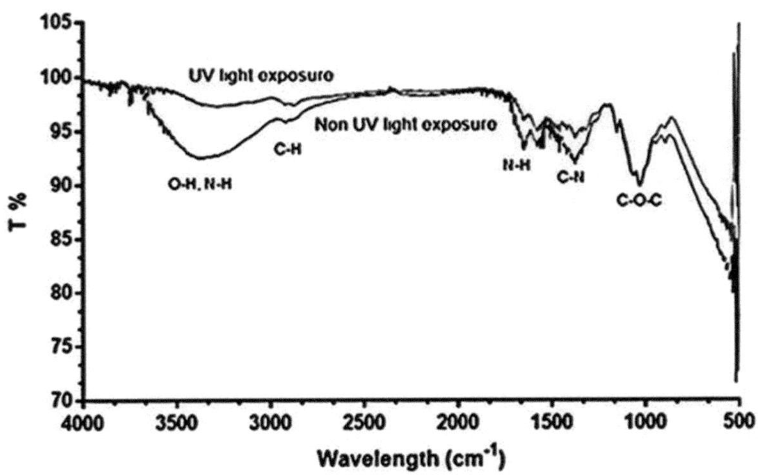

Figure 16.
FTIR spectrum of the 0.5% TiO_2 film after the UV exposure (blue line) for 24 h and without any exposure (black line).

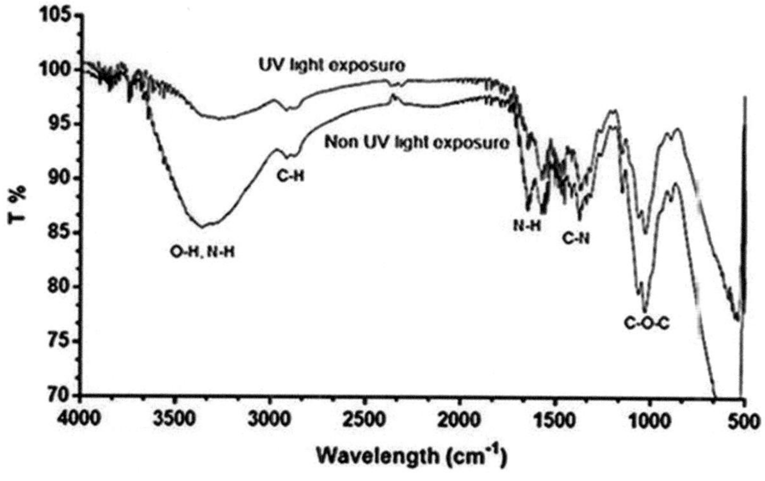

Figure 17.
FTIR spectrum of the 1.0% TiO_2 film after the UV exposure (blue line) for 24 h and without any exposure (black line).

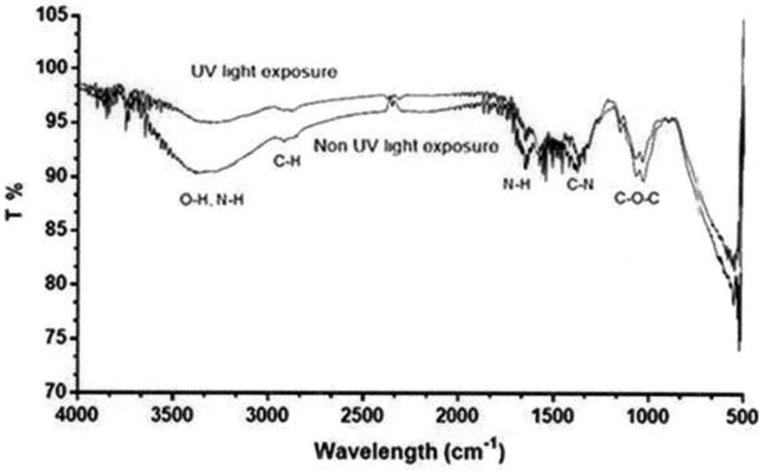

Figure 18.
FTIR spectrum of the 1.5% TiO_2 film after the UV exposure (blue line) for 24 h and without any exposure (black line).

almost perfectly with the anatase pattern. This is an important finding since any formation of the two other polymorphs or even the amorphous alternative would have resulted in a loss of the photocatalytic capacity of the dioxide.

6.5 Scanning electron microscopy (SEM)

Scanning electron microscopy (SEM) helped acquire information about the external morphology (texture) of the films [44]. In this research, SEM confirmed the adequate TiO_2 dispersion within the chitosan matrix. **Figure 20** shows the 0.0% and 0.5% films with two (2) different positions and angle for each sample. **Figure 21** reveals analogous information for the films 1.0 and 1.5%. As the amount of NPs increased, the uniformity was maintained.

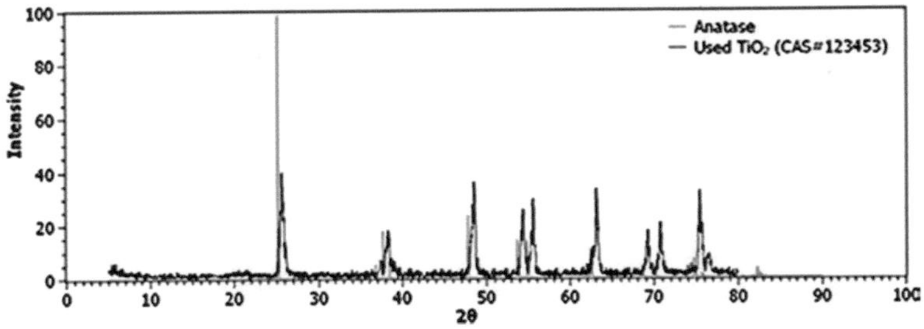

Figure 19.
XRD analysis diffraction of the TiO_2 used in the research, compared to pattern of anatase.

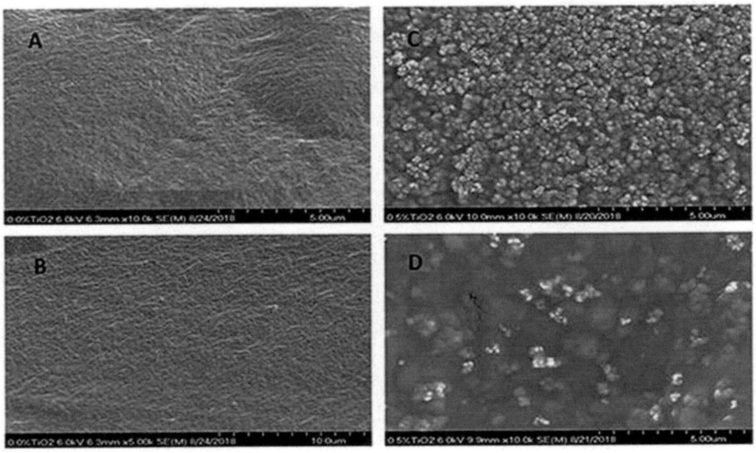

Figure 20.
SEM images of the films containing 0.0% TiO_2 (A, B) and 0.5% TiO_2 (C, D).

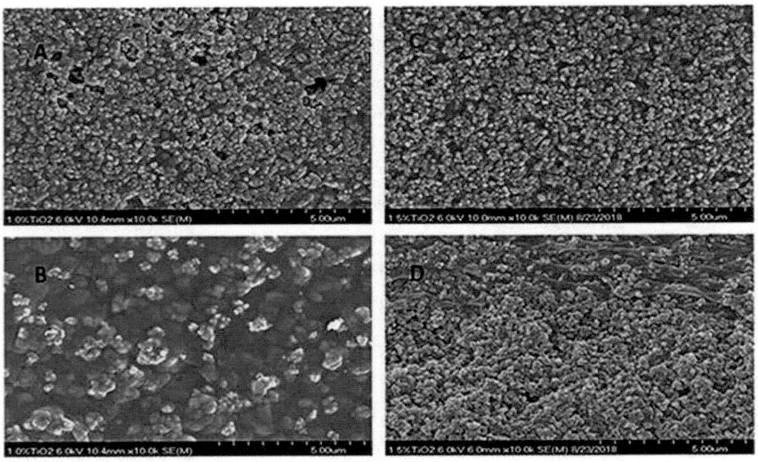

Figure 21.
SEM images of the films containing 1.0% TiO_2 (A, B) and 1.5% TiO_2 (D, C).

For each film, multiple shots were taken at different angles and positions within the films. **Figures 20** and **21** present the most typical configurations of the particle dispersion. **Figure 20** A and B shows the surface of the chitosan without NPs. SEM analysis in overall showed a uniform distribution of TiO_2 nanoparticle clots over the chitosan film as no large aggregate was observed. No phase separation was evident, as the TiO_2 content increased.

7. Discussion

As a whole, the results in Figures 5 through 12 prove clearly that the biocomposite film made of chitosan and TiO_2 NPs can be photoactivated. Once the films containing NPs have produced ROS, they are capable of reducing significantly the amount and growth rate of bacteria. Two different tests (viz., the growth curve analysis and the Kirby-Bauer technique) present incontrovertible evidence that when the bacteria are exposed to the photoactivated biocomposites, their growth pattern is affected, and the end amount is lower. We already mentioned that the reviewed literature proved that the main mechanism affecting the bacteria growth is the production of ROS rendered by the TiO_2 upon UV light irradiation. However, in order for these ROS to be effective (in sufficient concentration), the correct polymorph of TiO_2 is needed, i.e., anatase. For this reason, we selected this dioxide NPs, seeking to obtain more ROS, as mentioned in Section 4 of this chapter. Nonetheless, it is well-known that different processing conditions (involving temperature) of titanium dioxide can lead to a phase change. Thanks to the XRD characterization, we were able to confirm that the crystal structure used throughout this research was anatase and that the synthesis process of the biocomposite films did not change its crystal structure.

During the said process, the mixing methods used to disperse uniformly the TiO_2 NPs in the chitosan could have caused clots and agglomeration of the particles. This was the only reason to use a SEM (secondary imaging mode) to acquire the images shown in Section 6.5. The technique allowed scrutinizing the entire film exposed surface from different angles. This extensive survey did not evince any large clots or even lumps of NPs on the surface of the films. These images also proved that the attachment of the NPs onto the chitosan films was efficacious. In effect, as the TiO_2 content increased more, film surface was uniformly covered by the NPs, and no detachment occurred.

To reveal the interaction between TiO_2 and chitosan before and after UV irradiation, as already presented in Section 6.3, we acquired a set of FTIR spectra. Due to the addition of oxygen atoms from the TiO_2 NPs, one can observe a more intense peak from the O-H bond in **Figures 16–18**. Along that line, as the UV light photoactivated the TiO_2, the spectral data revealed a smaller intensity from this bond (O—H) after the UV irradiation. Finally, the bond nigh 3000 cm^{-1} (C—H) indicates a degradation of the chitosan polymer as this peak decreases This could result in a more brittle substrate to hold the nanoparticles, which should be taken into consideration in upcoming applications of the composite.

In closing, even with the rise of superbugs and the decline in effectiveness of antibiotics, innovated materials like nanoparticles show a promising future as the next generation of antimicrobial. In this context, our chitosan/TiO_2 biocomposite films can represent an economic option when compared to other materials for similar applications, such as activated carbon, capable of reducing the bacteria present in water. The synthesis process is reproducible and very economical, creating a technology with ample potential application.

8. Conclusions

This chapter presents an appealing alternative of a novel antibacterial material. The use of chitosan was due to its versatility, its abundance in nature, and excellent biocompatibility and nontoxicity. Through a straightforward technique, TiO_2 nanoparticles dispersed onto a chitosan matrix permitted formulating a renewable and biodegradable composite film capable of excellent antibacterial properties when exposed to UV light.

The synthesis process used did not affect the crystal structure of the dioxide, which remains anatase polymorph. X-ray diffraction characterization confirmed such anatase structure.

Chitosan demonstrated to be a very good selection as a matrix able to hold in place the TiO_2 nanoparticles. Incorporation of TiO_2 nanoparticles onto the chitosan film via solution casting method produced the chitosan/TiO_2 biocomposite films. The immobilized TiO_2 exhibited good distribution in the chitosan film, as demonstrated via secondary electron microscopy. From those observations, one can infer that chitosan provides an excellent substrate to immobilize TiO_2 nanoparticles. Electron microscopy also demonstrated that the nanostructured TiO_2 particles dispersed homogeneously within the polymer matrix without obvious aggregation even at the highest TiO_2 content, i.e., 1.5% wt.

Under UV light, TiO_2 proved effective against *E. coli* and *S. aureus*. This is due to the generation of ROS from the UV excitement of the TiO_2. The growth curve analysis demonstrated that the growth rate of bacteria was affected, as evidenced by the Kirby-Bauer technique. In the present experimental conditions, the best effect came from biocomposites containing 1.5% wt TiO_2.

The irradiation with UV light affected the chitosan covalent bonds, as revealed by the Fourier-transform infrared spectroscopy. From the spectra of chitosan with TiO_2, one could infer that more oxygen molecules were present. The degradation of the bond at 3000 cm^{-1} (C—H) demonstrated that extensive UV light irradiation is detrimental to the biopolymer, compromising its integrity in actual applications.

The research presented for the synthesis of the films is an effective way to immobilize TiO_2 while retaining the structure of the stronger photocatalytic polymorph.

Acknowledgements

This material is based upon work supported by the National Science Foundation under grants No. 0851449 and 1345156 (CREST Program). The authors would like to thank Ms. Barbara Sanchez along with the UPRM Biology Department for the use of their facilities. Professor Matias Cafaro provided insightful comments for the completion of this research.

Conflict of interest

The authors declare no conflict of interest.

Notes/thanks/other declarations

Most of the work completed in the laboratory was with the assistance of the following undergraduate students: Amanda Quintero, Yamalis Lopez, Christian McRobert, Katyria Torres, Kenneth Serrano, Karimar Amador, Ariana Torres, Luis Orta, Genesis Reyes, Luis Olmo, Karla Romero, and Edwin Ramos.

Author details

Johnny López Calero, Zuleika Oquendo Berríos and Oscar M. Suarez*
University of Puerto Rico at Mayagüez, Mayagüez, Puerto Rico

*Address all correspondence to: oscarmarcelo.suarez@upr.edu

References

[1] Hsueh P-R. New Delhi metallo-β-lactamase-1 (NDM1): An emerging threat among enterobacteriaceae. Journal of the Formosan Medical Association. 2010;**109**:685-687

[2] Coetzee J, Corcoran C, Prentice E, Moodley M, Mendelson M, Poirel L, et al. Emergence of plasmid-mediated colistin resistance (MCR-1) among *Escherichia coli* isolated from South African patients. South African Medical Journal. 2016;**106**(5):449-450

[3] Tsutsui M, Kawakubo H, Hayashida T, Fukuda K, Nakamura R, Takahashi T, et al. Comprehensive screening of genes resistant to an anticancer drug in esophageal squamous cell carcinoma. International Journal of Oncology. 2015;**47**(3):867-874

[4] Mehdipour Moghaddam MJ, Mirbagheri AA, Salehi Z, Habibzade SM. Prevalence of class 1 integrons and extended spectrum beta lactamases among multi-drug resistant *Escherichia coli* isolates from North of Iran. Iranian Biomedical Journal. 2015;**19**(4):233-239

[5] Knetsch MLW, Koole LH. New strategies in the development of antimicrobial coatings: The example of increasing usage of silver and silver nanoparticles. Polymers. 2011;**3**(1):340-366

[6] Poole K. Mechanisms of bacterial biocide and antibiotic resistance. Journal of Applied Microbiology. 2002;**92**(S1):55-64

[7] Jayaraman R. Antibiotic resistance: An overview of mechanisms and a paradigm shift. Current Science. 2009;**96**(11):1475-1484

[8] Gurunathan S, Han JW, Dayem AA, Eppakayala V, Kim JH. Oxidative stress-mediated antibacterial activity of graphene oxide and reduced graphene oxide in *Pseudomonas aeruginosa*. International Journal of Nanomedicine. 2012;7:5901-5914

[9] Zakharova OV, Godymchuk AY, Gusev AA, Gulchenko SI, Vasyukova IA, Kuznetsov DV. Considerable variation of antibacterial activity of Cu nanoparticles suspensions depending on the storage time, dispersive medium, and particle sizes. BioMed Research International. 2015;**2015**:412530, 1-11

[10] Leung YH, Ng AM, Xu X, Shen Z, Gethings LA, Wong MT, et al. Mechanisms of antibacterial activity of MgO: Non-ROS mediated toxicity of MgO nanoparticles towards *Escherichia coli*. Small. 2013;**10**(6):1171-1183

[11] Ramalingam B, Parandhaman T, Das SK. Antibacterial effects of biosynthesized silver nanoparticles on surface ultrastructure and nanomechanical properties of Gram-Negative bacteria viz. *Escherichia coli* and *Pseudomonas aeruginosa*. ACS Applied Materials & Interfaces. 2016;**8**(7):4963-4976

[12] Lim JY, Yoon J, Hovde CJ. A brief overview of *Escherichia coli* O157:H7 and its plasmid O157. Journal of Microbiology and Biotechnology. 2010;**20**(1):5-14

[13] Rasheed MU, Thajuddin N, Ahamed P, Teklemariam Z, Jamil K. Antimicrobial drug resistance in strains of *Escherichia coli* isolated from food sources. Revista do Instituto de Medicina Tropical de São Paulo. 2014;**56**(4):341-346

[14] Taylor T, Unakal C. *Staphylococcus aureus*. 2nd ed. Treasure Island, FL: StatPearls Publishing; 2018

[15] Bush L, Schmidt C, Perez M. Infecciones por *Staphylococcus aureus*—Infecciones—Manual

MSD versión para profesional. 2018. Available from https://www.msdmanuals.com/es/hogar/infecciones/infecciones-bacterianas/infecciones-por-staphylococcus-aureus; [Accessed: November 2018]

[16] Kodoth V, Jones M. The effects of ultraviolet light on *Escherichia coli*. Journal of Emerging Investigators. 2015;**102**(3):23-28

[17] Davis D. The History of Chitosan. 2018. Available from: https://www.streetdirectory.com/etoday/the-history-of-chitosan-euolup.html [Accessed: November 2018]

[18] Choi C, Nam JP, Nah JW. Application of chitosan and chitosan derivatives as biomaterials. Journal of Industrial and Engineering Chemistry. 2016;**33**:1-10

[19] Fernandez-Saiz P, Lagaron J, Ocio M. Optimization of the biocide properties of chitosan for its application in the design of active films of interest in the food area. Food Hydrocolloids. 2009;**23**(3):913-921

[20] Sedaghat F, Yousefzadi M, Toiserkani H, Najafipour S. Bioconversion of shrimp waste *Penaeus merguiensis* using lactic acid fermentation: An alternative procedure for chemical extraction of chitin and chitosan. International Journal of Biological Macromolecules. 2017;**104**:883-888

[21] Muxika A, Etxabide A, Uranga J, Guerrero P, de la Caba K. Chitosan as a bioactive polymer: Processing, properties and applications. International Journal of Biological Macromolecules. 2017;**105**(2):1358-1368

[22] Moura MJ, Brochado J, Gil MH, Figueiredo MM. In situ forming chitosan hydrogels: Preliminary evaluation of the in vivo inflammatory response. Materials Science and Engineering C. 2017;**75**:279-285

[23] Ogawa Y, Azuma K, Izawa H, Morimoto M, Ochi K, Osaki T, et al. Preparation and biocompatibility of a chitin nanofiber/gelatin composite film. International Journal of Biological Macromolecules. 2017;**104**(B):1882-1889

[24] Patil S, Nanduri L. Interaction of chitin/chitosan with salivary and other epithelial cells—an overview. International Journal of Biological Macromolecules. 2017;**104**(B):1398-1406

[25] Elshaarawy R, Mustafa F, van Geelen L, Abou-Taleb A, Tadros H, Kalscheuer R, et al. Mining marine shell wastes for polyelectrolyte chitosan anti-biofoulants: Fabrication of high-performance economic and ecofriendly anti-biofouling coatings. Carbohydrate Polymers. 2017;**172**:352-364

[26] Radtke A, Piszczek P, Topolski A, Lewandowska Ż, Talik E, Andersen IH, et al. The structure and the photocatalytic activity of titania based nanotube and nanofiber coatings. Applied Surface Science. 2016;**368**:165-172

[27] Kudo A, Miseki Y. Heterogeneous photocatalyst materials for water splitting. Chemical Society Reviews. 2009;**38**(1):253-278

[28] Polderman MC. New Applications of UVA-1 Cold Light Therapy. Doctoral thesis. Leiden University; 2006

[29] Zheng M, Mingyuan G, Yanping J, Guoliang J. Preparation, structure and properties of TiO_2–PVP hybrid films. Materials Science and Engineering B. 2000;**77**(1):55-59

[30] Sayes CM, Wahi R, Kurian PA, Liu Y, West JL, Ausman KD, et al. Correlating nanoscale titania structure with toxicity: A cytotoxicity and inflammatory response study with human dermal fibroblasts and human

lung epithelial cells. Toxicological Sciences. 2006;**92**(1):174-185

[31] Blinova I, Ivask A, Heinlaan M, Mortimer M, Kahru A. Ecotoxicity of nanoparticles of CuO and ZnO in natural water. Environmental Pollution. 2010;**158**(1):41-47

[32] Davis R. Synthesis and characterization of VPI-5-supported titania clusters. Chemistry of Materials. 1992;**4**(6):1410-1415

[33] Welte A, Waldauf C, Brabec C, Wellmann PJ. Application of optical absorbance for the investigation of electronic and structural properties of sol–gel processed TiO_2 films. Thin Solid Films. 2008;**516**(20):7256-7259

[34] Wang L, Hu C, Shao L. The antimicrobial activity of nanoparticles: Present situation and prospects for the future. International Journal of Nanomedicine. 2017;**12**:1227-1249

[35] Hashimoto K, Irie H, Fujishima A. TiO_2 photocatalysis: A historical overview and future prospects. Japanese Journal of Applied Physics. 2005;**44**:8269-8285

[36] Leung S. Exposure to titanium dioxide and other metallic oxide nanoparticles induces cytotoxicity on human neural cells and fibroblasts. International Journal of Nanomedicine. 2008;**3**(4):533-545

[37] Zhukova LV. Evidence for compression of *Escherichia coli* K12 cells under the effect of TiO_2 nanoparticles. ACS Applied Materials & Interfaces. 2015;**7**(49):27197-27205

[38] Jung WK, Koo HC, Kim KW, Shin S, Kim SH, Park YH. Antibacterial activity and mechanism of action of the silver ion in *Staphylococcus aureus* and *Escherichia coli*. Applied and Environmental Microbiology. 2008;**74**(7):2171-2178

[39] Pan X, Wang Y, Chen Z, Pan D, Cheng Y, Liu Z, et al. Investigation of antibacterial activity and related mechanism of a series of nano-$Mg(OH)_2$. ACS Applied Materials & Interfaces. 2013;**5**(3):1137-1142

[40] Sangari M, Umadevi M, Mayandi J, Pinheiro JP. Photocatalytic Degradation and Antimicrobial Applications of F-doped MWCNTs/TiO_2 Composites. Spectrochimica Acta Part A: Molecular and Biomolecular Spectroscopy. 2015;**139**:290-295

[41] Fourier Transform Infrared Spectroscopy (FT-IR) | NUANCE: Northwestern University Atomic and Nanoscale Characterization Experimental Center. 2018. Available from: http://www.nuance.northwestern.edu/keck-ii/keck-ii-instruments/ftir-spectroscopy/index.html [Accessed: November 2018]

[42] Causserand C, Aimar P. Characterization of filtration membranes. In: Comprehensive Membrane Science and Engineering. Oxford, UK: Elsevier. Vol. 1. 2010. pp. 311-335

[43] Bish DL, Post JE. Modern Powder Diffraction. Washington, DC: The Mineralogical Society of America; 2006

[44] Egerton RF. Physical Principles of Electron Microscopy: An Introduction to TEM, SEM, and AEM. US: Springer; 2005

Chapter 6

Development of Efficient Composites via Renewable, Recyclable, and Degradable Additives

Seyyed Alireza Hashemi, Seyyed Mojtaba Mousavi,
Ali Mohammad Shakibafard, Ghasem Shouli,
Yousef Amrollahi and Aziz Babapoor

Abstract

Rapid development of composite industries raised the demand for tough, effective, practical, and sustainable composites with enhanced mechanical, electrical, thermal, and physical properties. However, several major problems such as high production cost and nonrecyclability and nondegradability of fillers and composites limited their selectivity and approaches toward their aims. Herein, our aim is to present and review the recent achievements in matter of reinforced composites with renewable, recyclable, or degradable additives toward development of composites through a mass-scale production strategy with least charges, high efficiency, and fine mechanical, thermal, and physical properties. Additionally, such aim can reduce the overall amount of nondegradable pollutions such as plastics in the nature and decline overall the charges of composite industries by reusing such materials and decreasing their demand for raw material. Last but not least, to clean up the nature from industrial wastes, green strategies should be developed to whether reuse such material or degrade them via practical strategies.

Keywords: renewable resources, degradable materials, recyclable fillers, composites, nanocomposites

1. Introduction

Polymeric-based composites have been frequently used in last decades due to their promising properties and selectable specification. It is widely known that the usage of a polymer and one (or more) type of solid fillers allows us to obtain several advantages. In this case, by using diverse kinds of fillers, we can develop composite structures with selectable mechanical, thermal, electrical, and chemical properties. Among the most common fillers in the composite industry, calcium carbonate, glass fibers, talc, kaolin, mica, wollastonite, silica, graphite, synthetic fillers (e.g., PET- or PVA-based fibers), and high-performance fibers (carbon, aramid, etc.) can be mentioned [1]. However, the usage of such additives leads to one of the main limitations of polymer composites; the presence of two different components within the composite structure makes their recyclability and reuse a bit difficult to the level that it becomes much more economical to dispose them into the dump rather than reusing them [2–8]. Such method is often considered as an unappealing alternative due to its high cost, environmental impacts, and technical limitation/difficulties. Furthermore, it is worsened by the fact that plastic production requires a remarkable consumption of oil-based resources, which are notoriously nonrenewable [9]. Increased pressure from environmental activists, preservation of natural resources, and attended stringency of laws passed by developing countries lead to the invention and development of natural materials with a focus on renewable raw materials [10–12].

In case of chronological order, primary attempts were focused on the fabrication and characterization of polymeric composites based on recyclable polymers (e.g., polyolefin) filled with natural organic fillers, i.e., derived fibers and materials derived from natural plants. Several factors follow this choice. In this regard, the usage of natural organic fillers rather than traditional mineral inorganic additives leads to massive reduction in the usage of nonbiodegradable polymers and nonrenewable resources. Furthermore, these fillers are usually obtained from wastes or abundant plants; thus the final product becomes fairly economical and cheap. Moreover, these additives are less abrasive compared with inorganic mineral counterparts in matter of processing machinery. Likewise, these additives are also less dangerous for employees all of whom working in the production line due to their easy incineration and safe inhalation. The usage of such natural resources for reinforcement of composites can lead to the development of composite structures with lower density (or specific weight), ideal thermal properties, and fine acoustic insulation performance than with their mineral reinforced counterparts. Natural bio-fiber composites are emerging as a viable alternative to glass fiber reinforced plastics, especially in automotive applications. Thermoplastic biocomposites and biopolymer including PP, PE, PS, and PLA all of which were reinforced with natural fibers have attracted great attention toward themselves during the past few years. Such thermoplastic biocomposites, compared with those made using a thermoset matrix, can be processed into different shapes and have the potential to be mechanically recycled [13]. The synthesis of recyclable and degradable polymeric materials with a highly cross-linked three-dimensional network structure like PLA is very important for polymer science and technology for supporting a green sustainable society. PLA offers some advantages over the traditional nonbiodegradable polymers especially when their recycling is quite difficult or not economical. Likewise, there are also many restrictions due to the specification and properties of used materials, while a number of these challenges are expected to be resolved via blending PLA with other polymers, by making micro- and nanocomposites of PLA, coating with high barrier materials, and also through enhancement with practical polymeric materials [13].

The annual disposal of over 10 million tons of plastic wastes in both the EU countries and the USA has raised the demand for management and reuse of these nonbiodegradable waste streams. The synthetic polymeric materials containing displaced glasses, metals, wood, and ceramics have been widely used in many products, especially in the case of packaging. The commodity plastics, the so-called big four including PE, PP, PS, and PVC have been used in variety of forms such as films, rigid containers, and flexible bags which have revolutionized the packaging industry. However, disposal of these kinds of materials within the environment without the possibility of degradation led to rise of multitude environmental and ecological concerns. The important feature of composite materials is that they can be designed and tailored based on selectable properties to meet different requirements. Since biofibers are cheap and biodegradable, biocomposites from biofiber reinforced biodegradable polymers will render a contribution in the twenty-first century due to the serious environmental problem [14].

2. Green composites

When traditional nonbiodegradable polymeric matrices are reinforcing with natural fibers (e.g., green fiber and biofibers), the resulting composite structure is becoming partially biodegradable. If the polymeric matrix being selected forms biodegradable resins (i.e., green polymer or biopolymer) and reinforced with natural fibers, the final composite structure is considered as green or fully biodegradable composite. Two or more diverse natural fibers as additives in combination with polymeric matrices (biopolymer or petroleum-based polymer) lead to the development of hybrid green composites. The aim of hybrid composite development is to customize the properties and specifications of resulting green composites. Another subclassification of green composites can be based on the natural fillers and functional behavior of green composites. Depending upon the type of natural fillers, green composites can be classified as unidirectional and bi-directional continuous fiber green composites or discontinuous reinforcement composites (aligned or randomly oriented in the form of particulates, short fibers, and whiskers). According to the functional behavior, they can be classified as functionally graded and smart green composites. At the end of their life cycle, green composites can be easily disposed into the soil without deteriorating the environment by creating pollution sources. Many scientific evaluations have been developed to address the potential and specifications of green and partially biodegradable composites [15]. **Table 1** reviews some of the reported works in case of these kinds of degradable composites.

Some processing techniques were widely used for fabrication of composites, while among them extrusion, subsequent injection, and compression molding can be mentioned. It was rarely observed that the overall amount of fillers exceed 50–60 wt%; however, in some cases higher percentages about 70–80 wt% were reported (especially in the USA). Besides, in more excessive situations, the compression molding method is preferred rather than injection molding method. Some common flaws related to the processing of such materials are due to the hygroscopic and hydrophilic nature of these additives along with their low thermal resistance which restrict the production procedure to temperatures lower than 200°C. In fact,

Fiber	Matrix	Composition type	Ref.
Rice husk, wheat, husk, and coconut coir	Epoxy resin (LY556) and hardener (HY951)	Partially biodegradable	[16]
Banana	Polylactic acid	Green	[17]
Banana	Epoxy resin (LY556) and hardener (HY951)	Partially biodegradable	[18, 19]
Jute	Polypropylene	Partially biodegradable	[20]
Jute	Polylactic acid	Green	[21, 22]
Sisal	Polylactic acid	Green	[23]
Sisal	Polypropylene	Partially biodegradable	[24]
Flax	Starch	Green	[25]
Flax	Polylactic acid	Green	[26, 27]
Kenaf	Polylactic acid	Green	[28, 29]

Table 1.
List of important partially biodegradable/green composites [15].

among the most highlighted problems of this field, we can refer to hygroscopic and hydrophilic nature of additives. Such features can significantly affect the distribution of fillers within the matrix and the interfacial adhesion between fillers and matrix. Additionally, in case of injection molding and during the processing of composites, the presence of humidity throughout materials leads to the formation of water vapor which can cause several systematic problems in the following molding steps where a venting or drying system is not present. In general, the formation of water vapor and also various kinds of gases during the processing leads to the formation of voids within the developed composite structure, thereby deteriorating their mechanical properties [9].

So far, the biodegradability of Mater-Bi-based green composites has not been perfectly investigated. However, available scientific research regarding the biodegradability of this type of composite revealed that they become biodegradable after their burial within the soil [30–32]. Rutkowska et al. [33] studied the biodegradation of bi-based green composites in different natural environments and observed a complete biodegradation after 4 weeks, and conducted these investigations only in case of neat materials. While these investigations were conducted only in case of neat materials. In another study, Scaffaro et al. [34] have studied Mater-Bi/wood flour composite biodegradation in active sewage sludge, while their obtained results showed that the developed composites undergo biodegradation with higher weight loss rates than the neat Mater-Bi. This effect was attributed primarily to the obtained morphology and capability of wood fibers to act as support for the bacterial growth [34]. Huda et al. [35] investigated the properties of PLA/recycled cellulose composites prepared by extrusion and injection molding. Their outcome showed that an increase in the filler loading up to the 30wt% significantly improved the rigidity without affecting the crystallinity degree or thermal stability of the final product. In another research, Plackett et al. [36] prepared PLA/jute composites through a film stacking technique. Their obtained data showed that the addition of jute can lead to considerable enhancement in the tensile properties, while it can also increase the brittle fracture and deteriorate the impact strength of the final product [9].

3. Bio-based and biodegradable polymeric composites

When a biodegradable material (neat polymer, blended product, or composite) is obtained completely from renewable resources, we usually call it a green polymeric material. Biopolymers from renewable resources have attracted much attention in recent years due to their advantages to nature [37]. Renewable resources of polymeric materials offer a solution to maintain sustainable development of economically and ecologically attractive technologies. The innovations in the development of materials from biopolymers, preservation of fossil-based raw materials, complete biological degradability of polymeric materials, reduction in the volume of garbage and composability in the natural cycle, protection of the climate through the reduction of carbon dioxide released, as well as the applications of agricultural resources for the production of bio/green materials are among the reasons that attracted the public attention toward such materials [38–40].

Bio-based and biodegradable polymeric materials have a wide range of applications in diverse fields such as packaging, biomedical, and agricultural fields. The most common application of biodegradable materials is in case of blends for TPS and aliphatic/aromatic polyesters, such as PLA or polylactides, PCL, PBAT, and PHB [41]. Recently, PLA has attracted considerable attention toward itself due to its biodegradability in the composting situation. PLA has turned into an economically plastic material for industries which is commonly used for production of films and packaging materials, e.g., trays, bottles, and thin films for food packaging industry [42–44]. This cost affordable polymeric material can meet the requirement of industries for composite materials with ideal mechanical properties along with easy processability. However, PLA has some environmental impacts according to the life cycle assessment (LCA) [45]. PLA is a thermoplastic aliphatic polyester obtained from the ring-opening polymerization of lactide, which may be derived from the fermentation of sugar feedstock at competitive prices [14]. The same as most of thermoplastics, PLA can be molded into bottles, containers, and so on by injection molding, blow molding, etc. or be extruded into fibers, films, and sheets [13].

Polyester-derived by-products are playing an important role in case of biodegradable plastics due to their highly potential hydrolysable ester bonds. Polyester-derived materials are consisted from two main chemical groups, i.e., aliphatic (liner) and aromatic (aromatic rings) polyesters. Recently, various kinds of commercially biodegradable polyester materials were developed; among which we can refer to PHA, PHH, PHB, PHV, PLA, PCL, PBS, PBSA, AAC, PET, PBAT, and PTMAT [46].

The major constituents of biofibers (lignocelluloses) are cellulose, hemicellulose, and lignin. The overall amount of cellulose in lignocellulosic systems can vary based on the type and age of the plants or species that were used. Cellulose is a hydrophilic glucan polymer, while lignin is a phenolic compound, generally resistant to microbial degradation, but the pretreatment of fiber renders it susceptible to the cellulose enzyme [47, 48]. The exact chemical nature of the principal component of biofiber (the lignin) still remains obscure [14, 38]. The chemical compositions and structural parameters of some important biofibers are represented in **Table 2**. As shown in **Table 2**, specifications of natural fiber significantly vary from each other. Such variation may be due to the origin, age, retting (mode of extraction of fiber from the source) process adopted, etc. [14].

Type of fiber	Cellulose (wt%)	Lignin (wt%)	Hemicellulose (wt%)	Pectin (wt%)	Wax (wt%)	Microfibrillar/ spiral angle (deg.)	Moisture content (wt%)	Ref.
Bast	61–71.5	12–13	13.6–20.4	0.2	0.5	8.0	12.6	[49, 50]
Jute	71	2.2	18.6–20.6	2.3	1.7	10.0	10.0	[51, 52]
Flax	70.2–74.4	3.7–5.7	17.9–22.4	1.9	0.8	6.2	10.8	[49]
Hemp	68.6–76.2	0.6–0.7	13.1–16.7	1.9	0.3	7.5	8.0	[53]
Ramie Kenaf	31–39	15–19	21.5	—	—	—	—	[54]
Leaf	67–78	8–11	10–14.2	10	2	20	11	[55]
Sisal	67–78	8–11	10–14.2	10	2	20	11	[55]
PALF	70–82	5–12	—	—	—	14	11	[56]
Henequen	77.6	13.1	4–8	—	—	—	8	[56]
Seed	82.7	—	5.7	—	6	—	—	[57, 58]
Cotton	82.7	—	5.7	—	6	—	—	[57, 58]
Fruit coir	36–43	41–45	0.15–0.25	3–4	—	41–45	8	[57, 58]

Table 2.
Chemical composition and structural parameters of some natural fibers [14].

4. PLA bioplastics

PLA is one of the most studied bioplastics due to its recyclability and attractive specifications. PLA can be biodegraded under certain conditions, such as the presence of oxygen and moisture [59]. Moreover, PLA is a biodegradable material, and its usage within the biodegradable polymers has a high level of importance, because it significantly reduces the negative environmental impacts of wastes due to its biodegradability, improves the material recyclability, and eases monitoring external effects on the properties of the final composition. In this matter, the first reason is that the generation of wastes from industrial processes where the grinding process of the offcuts and blending with the neat polymeric matrix is integrated is unavoidable [60]. The second reason is the possibility of PLA wastes to be reused and recycled, because it is necessary to extend the lifetime service of PLA prior to discarding them into the biodisposal sites (i.e., composting sites) [61, 62]. Finally, the negligible environmental impacts of PLA could be ignored if it is not being possible for industries to recycle the obtained wastes [60].

In a work by Ohya et al. [63], they reported a procedure to develop graft polymerization of PLA on polysaccharides via TMS-protected polysaccharides. Based on their study, introduction of TMS protecting groups to chlorotrimethylsilane/pyridine can make the low molecular weight polysaccharides soluble in organic solvents and lead to an approach which can control the overall amount of initiating groups (i.e., the overall amount of graft chains). Moreover, via usage of trimethylsilyl protection method, PLA-grafted polysaccharides with variable lengths and number of graft chains were developed. The graft copolymer films present lower glass transition temperature (T_g), crystallinity, melting temperature, and higher viscosity than pure PLA film. What is more, the usefulness of graft copolymer as a plasticizer was investigated with 1:4 blend films prepared from the graft copolymers and PLA. The blend films showed lower T_g and crystallinity and higher viscosity than pure PLA film [64].

In another research, multiple extrusions of PLA up to ten times were investigated by Enkiewicz et al. [61] using a double screw extruder for granulation of PLA followed by laboratory injection molding press, for preparation of specimens. In this case, their obtained results showed that the tensile strength values of PLA did not significantly depend on the number of extrusion cycles, and they showed only a little diversity from each other, showing a slight total reduction of about 5.2% after ten extrusion processes (**Figure 1(a)**). Besides, a negligible decrease in the tensile strain was observed (about 2.2–2.4%) which did not depend on the number of extrusion process. On the other hand, by an increase in the number of preformed

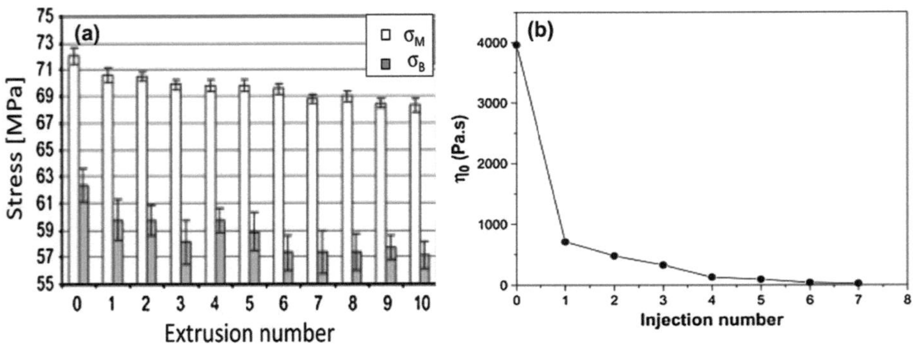

Figure 1.
(a) Tensile strength (σ_M) and tensile stress at break (σ_B) as functions of the extrusion number [61] and (b) zero viscosity of PLA as a function of injection number [60].

extrusions, the impact strength clearly declined (20.2% after ten extrusion processes), while the melt flow rate (MFR), water vapor, and oxygen transmission rates were significantly increased. The MFR value of the specimen subjected to ten extrusion processes was threefold higher than the original specimen. The thermal characteristics of specimens showed a slight decrease in the thermal stability of PLA, lower cold crystallization temperature, and a slight reduction in the melting point by increasing the number of cycles without any effect on the glass transition temperature (T_g) [61, 64].

PLA degradation has been found to be dependent on several factors, including molecular weight, purity, crystallinity, temperature, pH, the presence of terminal carboxyl or hydroxyl groups, water permeability, and additives all of which acting catalytically that may include enzymes, bacteria, or inorganic fillers [65]. Pyrolysis is a method which can lead to the treatment of polylactide, in which applied heat to a condensed substrate can lead to its chemical decomposition. Co-pyrolytic methods attracted considerable attention toward themselves due to their presented alternatives for disposal and convert wastes original form into valuable fuel sources. Likewise, the specific benefit of this procedure could be attributed to its potential for reduction of wastes within the nature, recovery of used chemical compounds, and replacing the developed fuel with frequently used fossil fuels. Furthermore, co-pyrolysis treatment of PLA-derived materials and biomass offers a potential alternative way for removal of wastes from the nature and could act as an upgrading approach during the pyrolysis of biomass and wastes [65]. PLA can be recycled to its primary form (i.e., monomer) through hydrolyze by boiling water or steam of lactic acid. This approach could lead to molecular recycling and provide an alternative for recycling both fabricated plant materials and post-consumed packages. In this regard, PLA could be hydrolyzed to its primary form by applying heat at 180–350°C for 30min, which can lead to recycle of L-lactic acid as the final outcome [66].

In a research by De Jong et al. [67], they depolymerized PLA through the release of dimers in alkaline conditions, in which the end chain degradation could be explained via intramolecular transesterifications. In this matter, an electrophilic attack which is catalyzed via the base of the hydroxyl (OH^-) end group on the second carbonyl group can lead to the formation of a ring and shorten the polymeric chain by hydrolysis of the final lactide. In the next step, free lactide hydrolysis into two lactic acid molecules. During this step, a random alkaline attacks to the carbon atoms of the ester groups, followed by the hydrolysis of the ester bond which leads to the intramolecular degradation of the primary chemical compound. Thereby, new low molecular weight molecules are produced; however, in the acidic conditions, the protonation process of the hydroxyl end group leads to an intramolecular hydrogen bond. The hydrolysis of the ester group can decrease the degree of polymerization of PLA due to the release of lactic acid molecules. Besides, random intramolecular protonation of ester groups' carbon atoms condenses the hydrolysis of ester bonds and produces diverse kinds of low molecular weight fragments [46]. Another study [60] on the reprocessing of PLA (containing 92% L-lactide and 8% D-lactide) showed that only the tensile modulus remains constant with thermomechanical cycles of up to seven injection moldings. Moreover, stress and strain at break, modulus, hardness, and rheological factors generally showed a small decline. Additionally, the viscosity of PLA decreased greatly (from 3960 to 713 Pa.s) after only one injection cycle. Zero viscosity (g_0) of PLA as a function of injection number is presented in **Figure 1(b)**. The mechanical properties of the recycled PLA had become too poor for an industrial application of the polymer [14]. In **Table 3**, a list of starchy and cellulosic materials that were used for the production of lactic acid can be seen.

Substrate	Microorganism	Lactic acid yield (g/l)
Wheat and rice bran	*Lactobacillus* sp.	129
Corn cob	*Rhizopus* sp. MK-96–1196	90
Pretreated wood	*Lactobacillus delbrueckii*	48–62
Cellulose	*Lactobacillus coryniformis* ssp. *torquens*	0.89
Barley	*Lactobacillus casei* NRRLB-441	0.87–0.98
Cassava bagasse	*L. delbrueckii* NCIM 2025 and *L. casei*	0.9–0.98
Wheat starch	*Lactococcus lactis* ssp. *lactis* ATCC 19435	0.77–1
Whole wheat	*Lactococcus lactis* and *Lactobacillus delbrueckii*	0.93–0.95
Potato starch	*Rhizopus oryzae* and *R. arrhizus*	0.87–0.97
Corn, rice, and wheat starches	*Lactobacillus amylovorus* ATCC 33620	<0.70
Corn starch	*L. amylovorus* NRRL B-4542	0.935

Table 3.
Starchy and cellulosic materials used for the production of lactic acid [46].

5. Nano-biocomposites

Nano-biocomposites are among the materials that contain bio-based polymers and low additions of nanoparticles of natural fibers such as cellulose and lignin for reinforcement purposes [68–70]. They are shaded orange in the bioplastics Spectrum because nanoparticles pose unknown hazards, and fibers are obtained via the craft method and can be treated with chemicals (isocyanates, alkalis) to enhance their properties as reinforcements [70]. Hazards of the craft method and these chemicals were described above. The health effects of nanoparticles are a major concern due to the lack of knowledge about their stability during processing, and there are potential toxicity concerns related to decomposition and/or migration during service [71]. Toxicologists hypothesize that nanoparticles may not be detected by the normal defense system of organisms; their small size can modify protein structures, and they can travel from respiratory system to the brain and other organs [72–74].

In a work by Hubbe et al. [75], they investigated about cellulose nanocrystals for their possible applications within the industry and stated that retention of developed nanocrystal properties should be controlled and guaranteed by using water miscible polymer matrices (e.g., starch products, latex, polyvinyl alcohol) to ease production procedure of cellulose nanocrystals and make them much more compatible with matrix. In another study, Eichhorn et al. [76] reported possible procedures of cellulose nanofiller recovery and then focused on the usage of cellulose nanowhiskers for the manufacturing of shape memory nanocomposites, as well as on the interfacial phenomena occurring in polymer/nanocellulose filler composites [9].

In another study by Mousavi et al. [77], NBR was composed with the natural polymers such as starch and glycerol. They also used silica nanoparticles to enhance the physical and mechanical properties of NBR. It was revealed that by the increase in the overall amount of starch, the mechanical properties of developed composite have considerably declined, while by an increase in the amount of starch, the module has increased, and impact resistance and also elongation at break have decreased. However, by the addition of silica nanoparticles, physical and mechanical properties of final composition were slightly improved, and the highest mechanical properties were achieved when silica nanoparticle filler loading increased to 3 wt%. Besides, the low weight percentage of starch can convert polymers into the biodegradable polymer which has a variety of applications in the automotive industry. In this matter, SEM examinations of specimens justified the biodegradability of reinforced composites by adding starch and silica nanoparticles to their system. SEM image of developed specimen can be seen in **Figure 2**.

Also recently, Mousavi et al. [3] developed polypropylene-wood composite as a new source of raw material with a combination of maleic anhydride and eggshell nanoparticles in order to improve their overall performance. They indicated that by creating cross-link between these additives, the tensile properties of polypropylene improve at high temperatures. The obtained results also showed that the tensile strength and fracture strain of samples containing cross-linked fillers and matrix are higher than samples without it at high temperatures. They also revealed that an optimum amount of peroxide is needed to achieve the desired tensile properties. Furthermore, their results showed that an increase in the overall amount of natural polymers and additives such as starch, corn, and eggshell nanoparticles as fillers can significantly enhance the physical and mechanical properties of developed composites. Their obtained results can be seen in **Tables 4–6** and **Figure 3**.

In a study by Farazi et al. [78], they prepared an antimicrobial substrate using LDPE/EVA/PE-MA/clay nanoparticle blend along with PS and GO. In this regard, LDPE was selected due to its fine elasticity, transparency, low melting point, and simple reversible process. Besides, developed matrix was well-reinforced with various kinds of natural, biodegradable, and antimicrobial additives. In this case, a twin screw extruder was used to produce the related specimens. Furthermore, specifications of developed samples were examined using mechanical, SEM, TEM, XRD, water absorption, water vapor, permeability, oxygen permeability, and microbial permeability tests. Moreover, the outcome of their study showed that developed specimens have appropriate mechanical properties along with their antimicrobial performance. Additionally, the outcome of water absorption, water vapor, and oxygen permeability tests showed that the sample containing GO/clay nanoparticles is presenting the best results. In this matter, the outcome of mechanical tests can be seen in **Figure 4(a–c)**, while the cultivating conditions of microorganisms and antimicrobial performance of developed samples can be seen to be negative on antimicrobial films (**Table 7**).

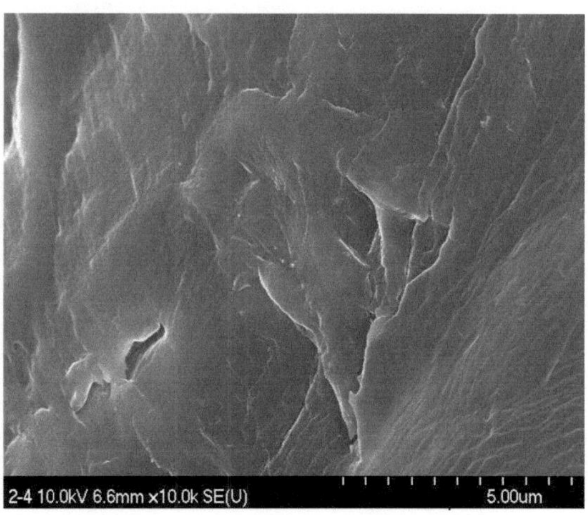

Figure 2.
SEM image of a specimen containing 5 wt% starch and 18.7 wt% NBR [77].

Sample number	Propylene (wt%)	Corn (wt%)	Starch (wt%)	Eggshell nanoparticles (wt%)	Compatibilizer (wt%)	Foaming agent (wt%)
1	100	0	0	0	0	0
2	50	46	0	0	4	0
3	50	44	0	0	4	2
4	50	33	10	1	4	2
5	50	21	20	3	4	2
6	50	9	30	5	4	2

Table 4.
Composition of developed specimens [3].

Sample number	Bending module (MPa)	Bending strength (MPa)	Tensile modulus (MPa)	Tensile strength (MPa)
1	522.57	7.249	6.08	17.42
2	1185.63	15.711	7.09	20.45
3	1224.27	15.908	6.68	19.8
4	1190.47	15.293	7.1	19.1
5	1025.55	13.713	1.66	18.49
6	925.1	12.478	12.41	19.59

Table 5.
Physical and mechanical properties of polypropylene-based composite [3].

Sample number	1	2	3	4	5	6
Impact strength (J/m)	17.55	30.86	14.88	51.43	222.9	55.28

The test conditions were as follow: Standard ISO 180, humidity about 45.1%, temperature about 22.2°C, touch back width of 8 mm, weights of 1, and touch radius of 0.25 mm [3].

Table 6.
Impact test of polypropylene composites based on ASTM 256 standard.

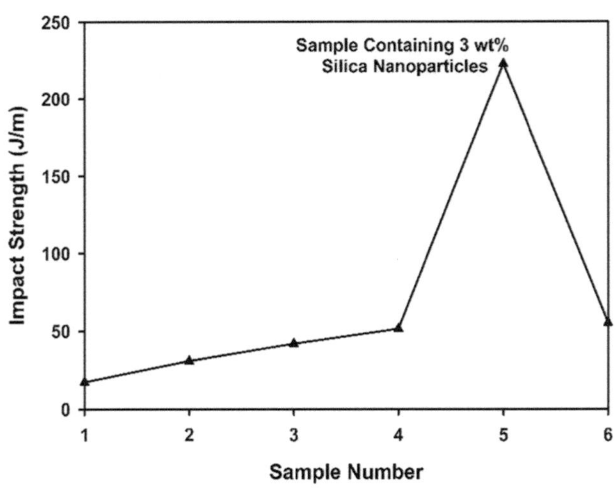

Figure 3.
Impact test results of polypropylene-based composites containing glycerol and starch. Sample 5 has the highest impact strength which is containing 3 wt% silica nanoparticles [3].

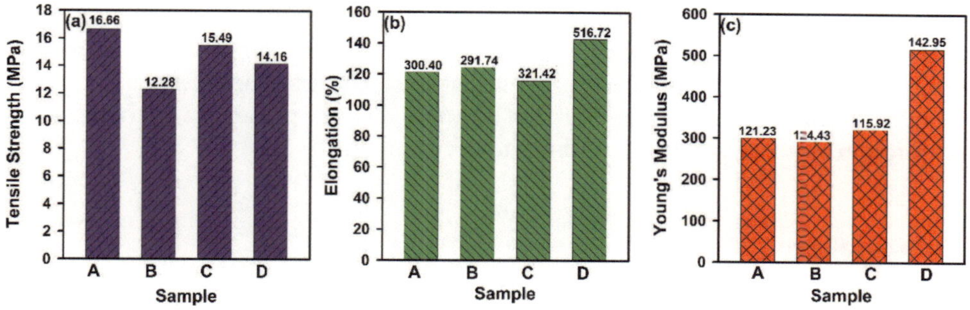

Figure 4.
(a) Tensile strength, (b) elongation at break, and (c) Young's modulus of developed samples by Farazi et al. Sample A (LDPE-EVA-PE-MA), sample B (LDPE-EVA-PE-MA-PS), sample C (LDPE-EVA-PE-MA-GO), and sample D (LDPE-EVA-PE-MA-PS-GO) [78].

Type of microorganism	ISIRI standard	Cultivating conditions
Plural	10,899-1ISIRI	20–25°C/3–5 days
Yeast	10,899-2ISIRI	20–25°C/3–5 days
E. coli	2946ISIRI/gram negative	37°C/17 h
Staphylococcus aureus	6806ISIR/gram negative	37°C/24 h
Salmonella	1810ISIRI/gram negative	37°C/24 h
Coliform	9263ISIRI/gram negative	37°C/24 h
Enterococcus	2198ISIRI/gram positive	35°C/24 h
Lactic acid bacteria (LAB)	13559ISIRI	30°C/72 h
Esporas Clostridium Sulfito reductor	9432ISIRI	37°C/48 h
Bacillus cereus	2324ISIRI/gram positive	30°C/18 h

Table 7.
Microorganism, medium, and cultivating conditions [78].

6. Conclusions

Green composites have attracted great attention toward themselves due to the ecological issues and decline of petroleum-based resources because of their hazards. Different types of natural fibers and their properties have been studied as a potential replacement for synthetic fibers. In recent years, various kinds of polymer composites were reinforced with organic fillers (rather than mineral inorganic additives) to reduce the usage of whether petroleum-based or mineral products and replace them with renewable resources. These "green" composites have numerous industrial applications. One of the main advantages of these materials is their low cost and degradability which protect the environment against nondegradable products. On the other hand, these composites are suffering from some disadvantages including ductility, processability, and dimensional stability. To resolve the aforementioned problems, researcher from all around the globe conducted great efforts to provide practical and appropriate alternatives through chemical treatment of additives and the usage of adhesion promoters to improve the interaction between matrix and fillers. Additionally, to obtain a completely biodegradable composite structure, it is bare bone essential to reinforce polymeric matrices with biodegradable additives rather than obtained additives form mineral or nonrenewable resources. In this regard, recent efforts were focused on the selection of biodegradable additives and matrices along with the optimization of their production procedures and processing parameters. Last but not least, the usage and development of sustainable and renewable resources not only can save our planet but also reduce production cost and pollutions within the environment.

Abbreviations

PET	polyethylene terephthalate
PVA	polyvinyl alcohol
PP	polypropylene
PE	polyethylene
PS	polystyrene
PLA	polylactic acid
PVC	polyvinyl chloride
TPS	thermoplastic starch
PCL	polycaprolactone
PBAT	polybutylene adipate terephthalate
PHB	polyhydroxybutyrate
PHA	polyhydroxyalkanoates
PHH	polyhydroxyhexanoate
PHV	polyhydroxyvalerate
EVA	ethylene-vinyl acetate
PBS	polybutylene succinate
GO	garlic oil
PALF	pineapple leaf fiber
TEM	transmission electron microscopy
SEM	scanning electron microscopy
XRD	X-ray diffraction
PBSA	polybutylene succinate adipate
LCA	life cycle assessment
AAC	aliphatic-aromatic copolyesters
PBAT	polybutylene adipate terephthalate
PTMAT	polymethylene adipate/terephthalate
TMS	trimethylsilyl
MFR	melt flow rate
NBR	nitrile-butadiene rubber
LDPE	low-density polyethylene
MA	maleic anhydride
PS	potassium sorbate

Author details

Seyyed Alireza Hashemi[1,2]*, Seyyed Mojtaba Mousavi[1,2], Ali Mohammad Shakibafard[1,2,3], Ghasem Shouli[3], Yousef Amrollahi[4] and Aziz Babapoor[5]

1 Department of Medical Nanotechnology, School of Advanced Medical Sciences and Technologies, Shiraz University of Medical Sciences, Shiraz, Iran

2 Pharmaceutical Sciences Research Center, Shiraz University of Medical Sciences, Shiraz, Iran

3 Zagros Petrochemical co. Asalouyeh, Buahehr, Iran

4 Borj va Baru Fars Research Center, Fars Science and Technology Park, Shiraz, Iran

5 Department of Chemical Engineering, University of Mohaghegh Ardabili, Ardabil, Iran

*Address all correspondence to: s.a.hashemi0@gmail.com

References

[1] Mousavi SM et al. Silica nanoparticles/phenol Novolac epoxy resin Nanocomposites via vacuum shock technique: Effect of silica nanoparticles on curing kinetic. Journal of Chemical, Biological and Physical Sciences. 2018;**8**:361-376

[2] Mousavi SM, et al. Investigation thermal stability and Activation energy composite of Epoxy phenol novelac Novalac/Unsaturated poly Ester/silica nano particle. 2015

[3] Mousavi SM et al. Modifying the properties of polypropylene-wood composite by natural polymers and eggshell Nano-particles. Polymers from Renewable Resources. 2015;**6**:157-174

[4] Hashemi SA et al. Electrified single-walled carbon nanotube/ epoxy nanocomposite via vacuum shock technique: Effect of alignment on electrical conductivity and electromagnetic interference shielding. Polymer Composites. 2018;**39**(S2):E1139-E1148

[5] Hashemi SA, Mousavi SM. Effect of bubble based degradation on the physical properties of single wall carbon nanotube/epoxy resin composite and new approach in bubbles reduction. Composites Part A: Applied Science and Manufacturing. 2016;**90**:457-469

[6] Mousavi S et al. Improved morphology and properties of Nanocomposites, linear low density polyethylene, ethylene-co-vinyl acetate and Nano clay particles by Electron beam. Polymers from Renewable Resources. 2016;**7**(4):135

[7] Hashemi SA et al. Lead oxide-decorated graphene oxide/epoxy composite towards X-ray radiation shielding. Radiation Physics and Chemistry. 2018;**146**:77-85

[8] Mousavi SM et al. Enhancement of rheological and mechanical properties of bitumen by Polythiophene doped with Nano Fe_3O_4. JOM. 2018;**7**(2):1-10

[9] La Mantia F, Morreale M. Green composites: A brief review. Composites Part A: Applied Science and Manufacturing. 2011;**42**(6):579-588

[10] Anandjiwala RD, Blouw S. Composites from bast fibres-prospects and potential in the changing market environment. Journal of Natural Fibers. 2007;**4**(2):91-109

[11] Wittig W. Einsatz von Naturfasern in Kfz-Bauteilen. Kunstoffe im Automobilbau. 1994:1-7

[12] Mousavi SM et al. Green synthesis of silver nanoparticles toward bio and medical applications: Review study. Artificial Cells, Nanomedicine, and Biotechnology. 2018;**46**:1-18

[13] Soroudi A, Jakubowicz I. Recycling of bioplastics, their blends and biocomposites: A review. European Polymer Journal. 2013;**49**(10):2839-2858

[14] Mohanty AK, Misra Ma, Hinrichsen G. Biofibres, biodegradable polymers and biocomposites: An overview. Macromolecular Materials and Engineering. 2000;**276**(1):1-24

[15] Singh JIP, Dhawan V, Singh S. Development and characterization of green composites: A review. Asia Pacific Journals. 2015;**2**(1)

[16] Dhawan V, Singh S, Singh I. Effect of natural fillers on mechanical properties of GFRP composites. Journal of Composites. 2013;**2013**:792620

[17] Sajna V, Mohanty S, Nayak SK. Hybrid green nanocomposites of

poly (lactic acid) reinforced with banana fibre and nanoclay. Journal of Reinforced Plastics and Composites. 2014;**33**(18):1717-1732

[18] Srinivasan V et al. Evaluation of mechanical and thermal properties of banana–flax based natural fibre composite. Materials & Design. 2014;**60**:620-627

[19] Uthayakumar M et al. Influence of red mud on the mechanical, damping and chemical resistance properties of banana/polyester hybrid composites. Materials & Design. 2014;**64**:270-279

[20] Haydaruzzaman et al. Fabrication and characterization of jute reinforced polypropylene composite: Effectiveness of coupling agents. Journal of Composite Materials. 2010;**44**(16):1945-1963

[21] George G et al. Thermal, calorimetric and crystallisation behaviour of polypropylene/jute yarn bio-composites fabricated by commingling technique. Composites Part A: Applied Science and Manufacturing. 2013;**48**:110-120

[22] Arao Y et al. Strength improvement in injection-molded jute-fiber-reinforced polylactide green-composites. Composites Part B: Engineering. 2015;**68**:200-206

[23] Fu W, Xu X, Wu H. Mechanical and biodegradable properties of L-lactide-grafted sisal fiber reinforced polylactide composites. Journal of Reinforced Plastics and Composites. 2014;**33**(22):2034-2045

[24] Bajpai PK, Singh I. Drilling behavior of sisal fiber-reinforced polypropylene composite laminates. Journal of Reinforced Plastics and Composites. 2013;**32**(20):1569-1576

[25] Ibrahim H et al. Characteristics of starch-based biodegradable composites reinforced with date palm and flax fibers. Carbohydrate Polymers. 2014;**101**:11-19

[26] Hallila T et al. Enzymatic pretreatment of seed flax-and polylactide-commingled nonwovens for composite processing. Journal of Thermoplastic Composite Materials. 2014;**27**(10):1387-1398

[27] Yuan Y, Guo M, Wang Y. Flax fibers as reinforcement in poly (lactic acid) biodegradable composites. In: Intelligent Computing and Information Science. Berlin, Heidelberg: Springer; 2011. pp. 547-553

[28] Ochi S. Mechanical properties of kenaf fibers and kenaf/PLA composites. Mechanics of Materials. 2008;**40**(4-5):446-452

[29] Tawakkal IS, Cran MJ, Bigger SW. Effect of kenaf fibre loading and thymol concentration on the mechanical and thermal properties of PLA/kenaf/thymol composites. Industrial Crops and Products. 2014;**61**:74-83

[30] Di Franco C et al. Degradation of polycaprolactone/starch blends and composites with sisal fibre. Polymer Degradation and Stability. 2004;**86**(1):95-103

[31] Bastioli C. Properties and applications of mater-Bi starch-based materials. Polymer Degradation and Stability. 1998;**59**(1-3):263-272

[32] Alvarez V, Ruseckaite R, Vazquez A. Degradation of sisal fibre/mater Bi-Y biocomposites buried in soil. Polymer Degradation and Stability. 2006;**91**(12):3156-3162

[33] Rutkowska M et al. Biodeterioration of mater-Bi Y class in compost with sewage sludge. Polish Journal of Environmental Studies. 2004;**13**(1):85-90

[34] Scaffaro R et al. Degradation of mater-Bi®/wood flour biocomposites in active sewage sludge. Polymer Degradation and Stability. 2009;**94**(8): 1220-1229

[35] Huda M et al. "Green" composites from recycled cellulose and poly (lactic acid): Physico-mechanical and morphological properties evaluation. Journal of Materials Science. 2005;**40**(16):4221-4229

[36] Plackett D et al. Biodegradable composites based on L-polylactide and jute fibres. Composites Science and Technology. 2003;**63**(9):1287-1296

[37] Kaplan D. Biopolymers from Renewable Resources: Macromolecular Systems-Materials Approach. Berlin: Springer-Verlag; 1998

[38] Dilnawaz F et al. Dual drug loaded superparamagnetic iron oxide nanoparticles for targeted cancer therapy. Biomaterials. 2010;**31**(13):3694-3706

[39] Rouilly A, Rigal L. Agro-materials: A bibliographic review. Journal of Macromolecular Science, Part C: Polymer Reviews. 2002;**42**(4):441-479

[40] Mousavi SM et al. Octadecyl amine functionalized Graphene oxide towards hydrophobic chemical resistant epoxy Nanocomposites. ChemistrySelect. 2018;**3**(25):7200-7207

[41] Bastioli C. Handbook of Biodegradable Polymers. United Kingdom: iSmithers Rapra Publishing; 2005

[42] Van de Velde K, Kiekens P. Biopolymers: Overview of several properties and consequences on their applications. Polymer Testing. 2002;**21**(4):433-442

[43] Shibata M et al. Biocomposites made from short abaca fiber and biodegradable polyesters. Macromolecular Materials and Engineering. 2003;**288**(1):35-43

[44] Meinander, K., et al. Polylactides-degradable polymers for fibres and films. in Macromolecular Symposia. San Francisco, CA, USA: Wiley Online Library; 1997

[45] Klöpffer W. Life-Cycle Based Methods for Sustainable Product Development. Springer; 2003

[46] Nampoothiri KM, Nair NR, John RP. An overview of the recent developments in polylactide (PLA) research. Bioresource Technology. 2010;**101**(22):8493-8501

[47] David C, Fornasier R. Utilization of waste cellulose. 7. Kinetic study of the enzymatic hydrolysis of spruce wood pretreated by sodium hypochlorite. Macromolecules. 1986;**19**(3):552-557

[48] Paillet M, Peguy A. New biodegradable films from exploded wood solutions. Journal of Applied Polymer Science. 1990;**40**(3-4):427-433

[49] Bledzki A, Reihmane S, Gassan J. Properties and modification methods for vegetable fibers for natural fiber composites. Journal of Applied Polymer Science. 1996;**59**(8):1329-1336

[50] Hon DN. Chemical modification of lignocellulosic materials: Old chemistry, new approaches. Polymer News. 1992;**17**(4):102

[51] Ugbolue S. Structure/property relationships in textile fibres. Textile Progress. 1990;**20**(4):1-43

[52] Kucharyov M. Properties and modification methods for vegetable fibers for natural fiber composites. Tekst Prom. Journal of Applied Polymer Sciences. 1993;**23**:8-9

[53] Wuppertal E. Die Textilen Rohstoffe. Frankfurt: Dr. Spohr-Verlag; 1981

[54] Narayan R. Emerging technologies for materials and chemicals from biomass. In: ACS Symposium Series 476. American Chemical Society; 1992

[55] Varma D, Varma M, Varma I. Coir fibers: Part I: Effect of physical and chemical treatments on properties. Textile Research Journal. 1984;**54**(12):827-832

[56] Mukherjee P, Satyanarayana K. Structure and properties of some vegetable fibres. Journal of Materials Science. 1986;**21**(1):51-56

[57] Bledzki A, Gassan J. Natural Fiber Reinforced Plastics. New York: Marcel Dekker, Inc; 1997. pp. 787-809

[58] Varma D, Varma M, Varma I. Thermal behaviour of coir fibres. Thermochimica Acta. 1986;**108**:199-210

[59] Helfenbein D. Development of Recycling of PLA Sheet into PLA Pellets without Undergoing Chemical Process. Atlanta, GA: Atlanta; 2011

[60] Pillin I et al. Effect of thermo-mechanical cycles on the physico-chemical properties of poly (lactic acid). Polymer Degradation and Stability. 2008;**93**(2):321-328

[61] Żenkiewicz M et al. Characterisation of multi-extruded poly (lactic acid). Polymer Testing. 2009;**28**(4):412-418

[62] Badia J et al. Material valorisation of amorphous polylactide. Influence of thermo-mechanical degradation on the morphology, segmental dynamics, thermal and mechanical performance. Polymer Degradation and Stability. 2012;**97**(4):670-678

[63] Ohya Y, Maruhashi S, Ouchi T. Preparation of poly (lactic acid)-grafted amylose through the trimethylsilyl protection method and its biodegradation. Macromolecular Chemistry and Physics. 1998;**199**(9):2017-2022

[64] Ouchi T, Kontani T, Ohya Y. Modification of polylactide upon physical properties by solution-cast blends from polylactide and polylactide-grafted dextran. Polymer. 2003;**44**(14):3927-3933

[65] Park K, Xanthos M. A study on the degradation of polylactic acid in the presence of phosphonium ionic liquids. Polymer Degradation and Stability. 2009;**94**(5):834-844

[66] Gang W, Aimin L. Thermal decomposition and kinetics of mixtures of polylactic acid and biomass during copyrolysis. Chinese Journal of Chemical Engineering. 2008;**16**(6):929-933

[67] De Jong S et al. New insights into the hydrolytic degradation of poly (lactic acid): Participation of the alcohol terminus. Polymer. 2001;**42**(7):2795-2802

[68] Ahmann D, Dorgan JR. Bioengineering for pollution prevention through development of biobased energy and materials state of the science report. Industrial Biotechnology. 2007;**3**(3):218-259

[69] Lagaron JM, Lopez-Rubio A. Nanotechnology for bioplastics: Opportunities, challenges and strategies. Trends in Food Science & Technology. 2011;**22**(11):611-617

[70] Vilaplana F, Strömberg E, Karlsson S. Environmental and resource aspects of sustainable biocomposites. Polymer Degradation and Stability. 2010;**95**(11):2147-2161

[71] Arias JL et al. Development of iron/ethyl cellulose (core/shell) nanoparticles loaded with diclofenac sodium for arthritis treatment. International Journal of Pharmaceutics. 2009;**382**(1-2):270-276

[72] Álvarez-Chávez C et al. Sustainability of bio-based plastics: General comparative analysis. Journal of Cleaner Production. 2012;**23**:47-56

[73] Álvarez-Chávez CR et al. Sustainability of bio-based plastics: General comparative analysis and recommendations for improvement. Journal of Cleaner Production. 2012;**23**(1):47-56

[74] Mousavi SM et al. Development of Clay Nanoparticles toward Bio and Medical Applications. Bioresources. Intech open; 2018. pp. 167-191

[75] Hubbe MA et al. Cellulosic nanocomposites: A review. BioResources. 2008;**3**(3):929-980

[76] Eichhorn SJ et al. Current international research into cellulose nanofibres and nanocomposites. Journal of Materials Science. 2010;**45**(1):1

[77] Mousavi S et al. Investigation of physical, mechanical and biodegradation properties of nitrile butadiene rubber by natural polymers and Nano-silica particles. American International Journal of Research in Formal, Applied and Natural Sciences. 2014;**5**:110-117

[78] Farazi Z, Abdolrasoul O, Mojtaba MS, Alireza HS. Preparation of LDPE/EVA/PE-MA, Nano clay blend composite in the stage potassium Sorbate (KS) and garlic oil (GO) as an antimicrobial substance. Polymer Sciences. 2018;**4**:1-12

Chapter 7

Improving the Mechanical Properties of Natural Fiber Composites for Structural and Biomedical Applications

Owonubi J. Shesan, Agwuncha C. Stephen, Anusionwu G. Chioma, Revaprasadu Neerish and Sadiku E. Rotimi

Abstract

Natural fiber composites are designed for different purposes including structural and non-structural ones. These natural fiber composites vary greatly in their properties including mechanical properties. Mechanical properties which include the tensile and flexural properties are highly dependent on factors such as matrix type, filler type, processing, post processing treatment and many more, factors which are quite application specific. However, many research works develop their natural fiber composite before considering the possible applications. This chapter intends to X-ray the factors that affect the mechanical properties as it relates to structural and biomedical applications and suggest ways of improving the mechanical properties.

Keywords: natural composites, mechanical properties, structural applications, property improvement

1. Introduction

In recent times, the application of natural fiber polymer composites (NFPCs) for structural purposes has increased [1–7]. In the past, carbon fibers and basalt fibers reinforced polymer composites were commonly used due to their high performance to cost and high strength to weight ratios [8–15]. However, due to fundamental concerns, such as high cost of materials, durability, bonding integrity and gap in the development of standards for practice of abrasing or wearing of machines among others, their uses have been limited [16–19]. More importantly is the environmental concerns, which has strengthened the call for complete replacement of all synthetic fibers. Hence, the increase in research output for possible replacement of synthetic fibers with natural fibers in polymer composites [20–22].

Structural materials required for structural applications encompass materials that are primarily for the purpose of stress transmission or support. Structural materials may be in transportation (aircraft and automobiles), construction (building and roads), or in components used for the purpose of protecting the body (helmets and body armor), energy production (turbine blades), etc. [23–25]. Also, those used in microelectronics, are designed to meet specific performance requirements. Structural requirements may vary based on end use; hence the requirement for a material to be used for vehicle part may differ widely from what is required for manufacturing casing for electronic equipment. Generally, of more interest are the following properties: mechanical, thermal, electrical conductivity, dimension stability, water absorbability, etc. [26–31]. This work intends to critically review the mechanical requirements and suggest ways of improving the mentioned properties for structural and biomedical applications.

2. Mechanical properties of NFPCs

NFPCs are different depending on the type of polymers or fibers employed, the processing condition and the characteristic properties developed after processing [32, 33]. The use of these NFPCs for structural applications depends more on the mechanical properties. There are different mechanical evaluations carried out on a material and some are very specific to individual application. However, the general focus of most researchers is on a few tests including tensile test, flexural test, impact test, compression test and hardness. These few tests can describe the mechanical behavior of any material to a large extent and also gives an insight to other possible areas of applications.

2.1 Tensile properties

Tensile test is one of the fundamental and common types of mechanical test widely used for testing polymeric materials. It is also known as tension testing; it is used to determine the stress-strain behavior under tension [32]. In tensile testing, tensile force (pulling force) is applied to a material and the specimens' response to that applied force (stress) is measured. Under this test condition, samples are subjected to controlled tension until failure occurs. Tensile test determines how strong a material is and how long it can be stretched. Important information that can be obtained from tensile test includes; Young's modulus, yield strength, percentage elongation and ultimate tensile strength (UTS).

2.2 Flexural properties

Flexural test is used to determine the capability of a material to resist bending forces applied perpendicularly to its longitudinal axis and often called the transverse beam test. Flexural test properties are among the major parameters used in assessing the suitability of composite materials for structural applications. Parameters such as flexural load, flexural Young's modulus, flexural strength and deflection at break are measured and used to interpret the mechanical behavior under flexural stress.

2.3 Impact properties

Impact test is employed in assessing the impact strength, toughness and notch sensitivity of structural materials. In summary it is the capability of the material in question to withstand high rate loading. Toughness is the total of energy absorbed per unit volume of material before rupturing. The toughness is a measure of balance between strength and ductility of the material. Impact test is very critical for most polymer materials because it relates to the product performance and service life. It also influences other properties related to product safety and liability.

2.4 Compressive properties

Compressive tests evaluate the materials behavior when subjected to uniaxial compression load at a relatively low and uniform loading rate. These tests are very important for product design analysis, especially materials for building purposes. Compressive strength and compressive Young's modulus are the two major properties that are used; however, there are also compressive strain, deformation beyond yield point, and compressive yield stress.

2.5 Hardness properties

Hardness is that property of an engineering material which enables it to resist scratching, indentation, penetration and plastic deformation. It is a very significant property for any structural material from the engineering point of view, because hardness generally leads to increase in wear resistance by either erosion by water, oil and steam or friction.

2.6 Crashworthiness

This is the degree to which a vehicle or aircraft will protect its occupant from the effect of an accident [34]. It is the most important concept in vehicle defect cases. That is, the ability to prevent injuries to the occupant in the event of collision. Therefore, crashworthiness focuses on occupant protection to reduce fatality in the case of an accident. Different criteria can be used to determine crashworthiness depending on the nature of the impact and vehicle involved [35].

In general for composites, these properties depend on other factors such as fiber length, fiber weight percentage content and the extent of polymer/fiber interactions as will be discussed later.

3. Natural fiber polymer composites in structural purposes

Natural fiber reinforced polymer composites are attracting widespread interest for purposes which are sensitive to the materials weight, because their strength and stiffness combine well with their low density, however, their toughness is a major concern [36]. NFPCs are prepared using plant fiber as fillers, which are hydrophilic in nature and do not interact well with hydrophobic polymer matrix [37–39]. NFPCs are a set of important materials developed for numerous areas application; medical, pharmaceutical, food packaging, electronics, aerospace, automobile, construction, building, transport and many more [40–43]. This is because of the many unique qualities these materials possess or can be designed to possess. Qualities which include, but are not limited to; light weight, resistant to chemical attraction, resistant to corrosion, ability to be molded to any shape, can be processed using existing technology, environmental friendliness and sustainability. That is why the interest in these materials has grown tremendously in the last two decades [36, 44–46].

Dweib et al. fabricated bio-based composites for roof structures in the form of paper sheets, entirely from cellulose fibers and soy oil-based resins [47]. These developed sheets were tested for structural unit beams and were established to have given the necessary strength and stiffness for consideration in roof construction. Also Bektas et al. manufactured panels with a density of 0.7 g/cm^3 with a sunflower stalks percentage of 25, 50 and 75% fiber contents [48]. From the results of the mechanical tests, the panels were observed to have the required properties as required for general purpose-use particle board by normal standards.

Wood fibers/plastic composites have been used in large quantities for applications in window and door frames, decks, docks and molded panel components [49–51]. Natural fiber composites have been used to replace asbestos in the building industry, because of their health related issues [52]. The European Union policy tagged "end of life vehicle (ELV)" regulations, promulgated in 2003 and amended in 2005 and 2010, projected reduction of the final waste to be disposed at the end of life of vehicles to 5% by the year 2015. In the stated regulations, 85% of material used in manufacture of the vehicle must be recoverable through reuse or through recycling mechanically [49, 53]. This has generally increased interest and widespread use of NFPCs worldwide. It is noteworthy that this policy was promoted basically because of environmental and social concerns and not necessarily because of economic or technological reasons.

NFPCs durability and the availability of technology has allowed for large and complex shaped manufacturing of NFPCs, making them appealing in the automobile industries [54–56]. In Brazil, automobile industries consume, on average, 10–12.7 kg of natural fiber reinforcement per vehicle. These are circulated through the vehicle, such as rear door liners, front doors, boot liners, parcel shelves, sun roof interior shields and headrests [54]. Although NFPCs have gained tremendous interest in the industries, their applications are not unconnected to their environmental sustainability, low cost and renewability [57–60]. According to a review by Kiruthika [61], the challenge of replacing synthetic fibers completely in widespread applications is far from being overcome, with the improvement to the mechanical properties of composites being the major challenge.

4. Factors influencing the mechanical properties of NFPCs

NFPCs are prepared by compounding polymer matrix (either pristine or blend) with natural plant fibers. The fibers can be single or hybrid, microcellulose or nanocellulose. The natural fibers are made up of different chemical constituents and are exposed to different physical and chemical treatments, therefore, the properties of the resulting composites varied widely. Factors influencing their eventual composite(s) include the following:

4.1 Fiber type

Plant fibers are categorized based on the parts of the plant they are extracted from. Fibers can be extracted from the seeds, leaves or bast of the plants. Bast fiber is collected from the "inner bark" or the surrounding of the stem of certain dicotyledonous plants [61], like banana, flax, hemp, jute kenaf and ramie. These fibers have higher tensile strength and are mostly used in the packaging and paper industries [62, 63]. Sisal, pineapple and many others are extracted from the plant leaves while coir, cotton and abaca are examples of fibers extracted from plant seed. Generally, plant fibers give higher strength and stiffness; however their properties depend mainly on their structure and chemical composition. These invariably relate to the source of fibers, method of extraction, maturity, growing conditions, harvesting period, degree of retting and modification [64–67].

4.2 Fiber length, orientation and weight percentage loading

NFPCs properties are affected by the length of the fibers used, their distribution, the percentage of the fiber volume or volume fraction and their orientation within the matrix. For polymer composites, stress is transferred by the matrix through the fibers both at the interface along the fiber length and at the ends of the fibers by shear [68, 69]. Hence, the degree of load conveyed from the matrix to the fibers is a function of: (i) fiber length, which is referred to as critical fiber length or aspect ratio, (ii) orientation of the fibers and their direction relative to each other. If the fiber orientation and direction is not in the line of the applied stress, failure is bound to occur easily. Unidirectional fiber composites tend to transmit external stress better, that is why hand laid fiber composites performs better mechanically [70, 71].

Depending on the fibers in the matrix's orientation and direction, we can obtain three different types of reinforcement which include (i) longitudinally aligned fiber-filled composites, (ii) transversely aligned fiber-filled composites and (iii) randomly oriented short fiber composites [68, 72, 73]. While the longitudinally filled composites have low compression strength due to buckling of fibers and high tensile strength, the transversely filled composites on the other hand have low tensile strength. However, in the randomly oriented composites, it is far more difficult to predict the mechanical properties, due to the complexities of the distribution of load along the interface of the fiber and the matrix. Hence, considerable control over such elements as orientation dispersion and aspect ratio of the fibers, considerable improvement in the mechanical properties of the composites can be attained.

In general, high performance NFPCs can be obtained mainly by using materials with high fiber content, hence the effect of fiber loading on the properties of the NFPCs is of great significance. Also it is noteworthy to mention that additional fiber content of the composites invariably causes to increased tensile properties [74, 75].

4.3 Fiber-matrix adhesion

The effect of fiber-matrix adhesion cannot be over emphasized. A good number of researchers have reported their experimental results on the effect or importance of a good and strong fiber-matrix adhesion in the fabrication of composites with good mechanical properties [76, 77]. The type of bonds existing at the fiber-matrix interface greatly influences the mechanical properties of any fabricated composites. For a good transmission of stress from the matrix to the fiber to occur, the bond existing among the two components must be strong [64]. Due to the hydrophilic nature of the natural fibers, the interaction between the fibers and the hydrophobic polymer matrix is very weak [4, 78, 79]. Hence, the need to modify the fibers and introduce organic moiety that makes them more hydrophobic is necessary.

Fiber-matrix interface has been described as the reaction zone which plays a significant role in characterizing the composites mechanical properties [80]. A poor interaction between the two surfaces leads to poor transmission of load and therefore poor mechanical performance [81–83]. In addition, plant fibers need chemical modification for the distention or enlargement of the crystalline region, removal of surface impurities and elimination of hydrophilic hydroxyl groups for improvement to some of its relative properties [68].

4.4 Choice of polymer matrix

Polymer matrix could be either a thermoset or a thermoplastic, with varying preparation procedures and conditions, the performances of polymer matrix are affected quite differently. Thermoset are made in such a way that they develop good bonding with the fibers, especially during curing stage. However, the processes involved are time and energy consuming. Although in the case of tensile loading of the composites, the significance of matrix is evident, some researchers have reported good improvement with the same fiber when the matrices are changed. However, for compressive, in-plane shear and inter-laminar strength, they are highly influenced by the type of matrix used [84].

4.5 Processing conditions

The properties of NFPCS have been shown to vary from one processing technique to another [85–89]. Common techniques for the preparation of NFPCs include injection, extrusion, compression and resin transfer molding. These techniques use different processing conditions or parameters even when the materials being processed are the same. Changes in factors such as mixing speed, pressure and temperature can change the properties of the final product with any slight change [90]. For example, the preparation of sisal fiber polyester composites by employing both the compression molding and resin transfer molding technique (RTM) gave products with varying mechanical properties. The products of the RTM gave a composite with higher Young's modulus, tensile strength and flexural strength than the product of the compression molding [90]. Vacuum molding technique is one the simplest manufacturing methods for plastic materials [91]. It is suitably adapted for molding a required shape from a plastic sheet material. In this molding technique, a plastic sheet is heated up to its molding temperature using electric heat; it is then transferred to a molded shape. To obtain the shape, a vacuum is created between the mold and the sheet. Vacuum molding is an inexpensive method when compared to other molding methods. It has numerous application including, aircraft, skin tight packing, disposable tray and caps. It is a low cost methods already being employed in many areas of endeavors as mentioned. A lot of research work is being carried out on how to improve the vacuum forming method, for instance, vacuum-assisted resin transfer (VARI), also known as vacuum infusion process (VIP) which was reported by [92] to be considered as an attractive method for the production of NFPCs low cost and good performance. It uses low-cost one-sided tooling and injects low-viscosity resin into dry fiber that was performed under low pressure. The method was found to be economically suitable for the manufacturing of large composite structures, such as boat hull, wind turbine blade and aircraft structures with low or high volumes of production.

However for vacuum forming method the cycle time of production is still much affect by the cool rate and time, amount of pressure applied and the fiber content [93]. These factors can greatly affect the mechanical properties the final product.

4.6 Presence of void

The introduction of fibers into the matrix during processing is accompanied with the introduction of air and other volatile substances. These substances which are mostly trapped in the fibers may form voids in the composites after processing and curing along the individual fibers. This can negatively affect the composites mechanical properties. In addition, the rate of cooling during processing can also result to the formation of voids [73]. When the void content is too high, it leads to greater affinity for water diffusion, lowering fatigue resistances and increased disparity in mechanical properties [94].

4.7 Thermal stability

The importance of the stability in the preparation of the composites cannot be over emphasized as it affects the mechanical properties considerably. The different components of plant fibers are sensitive to different range of temperatures, i.e., hemicellulose, cellulose, pectin [95, 96]. Most fibers start degrading thermally at 220°C, thereby limiting the composites thermally [68]. In recent reports, the thermal stability of these fibers were greatly improved by removing maximally the lignin, hemicellulose and other alkaline soluble substances in the fiber through physical, chemical or biological means [97].

In general, the above mentioned factors do not influence the mechanical properties of NFPCs individually or in isolation. Rather, in the fabrication of a composite, the cumulative effect of two or more of these factors may be responsible for the composites mechanical failures. Jain et al. evaluated the effect of inter-fiber interactions of different class of reinforced polymer composites on the mechanical properties and the relation to stress field [98]. The research findings highlighted the important role of microstructural arrangement in the determination of the final response of the composites. They reportedly concluded that the local fiber matrix arrangement and their neighborhood density are highly influenced and sensitive to the stress and overall strain energy.

5. Methods for improving mechanical properties of NFPCs

There are different methods which can be employed for mechanical properties improvement of NFPCs. These methods are not rigid formulas that once applied will result in massive enhancements, rather improving mechanical properties of NFPCs is an active area which is ongoing. Mechanical improvement in one area of application may not necessarily yield the same result in another area of application. For example, improving stiffness of a material might be good for the construction and building industries but may not be required or worthwhile for biomedical applications, as it might prefer improvement in flexibility and toughness. Therefore all possible options will be provided; it is left for the researchers to select as appropriate. Furthermore, these methods can be combined to produce synergistic effect.

5.1 Surface modification

There are many literatures and reviews on the benefits of surface modification of natural fibers to the enhancement of fibers mechanical properties and by extension, the composites [99–101]. The mechanical properties of plant fibers depend greatly on the chemical structure, chemical composition and the structural arrangement of cellular fibrils [101]. Other factors such as climatic conditions, age, extraction procedures, growth condition and time of harvest also influence the mechanical properties of natural fibers. All these affect the percentage composition of cellulose in the fibers. Furthermore, the hemicelluloses and lignin are less thermally stable compared to cellulose, modification is one way of reducing the percentage content of hemicelluloses and lignin or even eliminating them completely.

Graphical sample of cellulose structure which consists of amorphous (untreated) and crystalline (treated) regions is shown in **Figure 1** [69]. There exist strong intra-molecular hydrogen bonds with large molecules in the crystalline regions of the cellulose ensuring the crystalline region is very compact and this makes it difficult for chemical penetration. On the other hand, the amorphous region is loose and allows penetration for possible modification.

Surface modification of plant fibers involves the treatment given to the plant fibers in order to increase its cellulose content, improve its interaction with the polymer matrix and also improve their mechanical, thermal and dimensional stability properties. Surface modification can be physical, chemical or biological [11, 21, 79, 102, 103].

Chemical modification involves chemical reaction with the fiber components, thereby making them soluble so that they can be removed by repeated washing. Alkali treatment also known as mercerization involve the use of alkali solution to dissolve all soluble contents of the fibers including wax, oil, pectin, lignin and some part of hemicelluloses [11, 104–107]. Alkali treatment makes the fibers surface rougher and reduce the fibers to fibrils [108, 109]. It improves the aspect ratio and

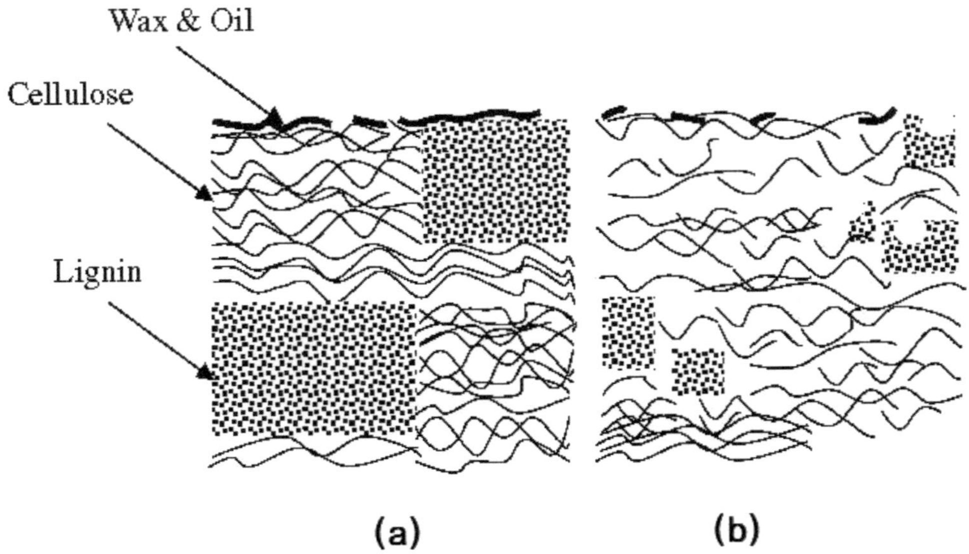

Figure 1.
Schematic diagram showing the crystalline (a)/amorphous (b) regions in a fiber material [69].

creates room for mechanical interlocking in composites [23, 110–112]. To improve chemical interactions in NFPCs which may lead to improved mechanical properties, more chemical treatments are required other than alkali treatment. Another advantage of alkali treatment is that it exposes the OH group on the cellulose chains which can again be utilized for further chemical modification in some cases and in others it can allow for chemical treatment to introduce functional groups onto the surfaces of the fibers. These functional groups can then be used for chemical interaction during the preparation of the composites [27, 113, 114]. Such chemical treatments include silane, sodium silicate, oxidization treatments with DCP, or $KMnO_4$ treatments.

Biological methods involve the degradation by biological organism of all other components of the fiber except the cellulose. This is because the cellulose is highly crystalline and so cannot be easily degraded. Biological treatments are more environmentally friendly, produces more crystalline cellulose with better mechanical and thermal properties and it is cost effective. However, it is laborious and time consuming [115–121].

Physical methods of surface modification involve the use of plasma and corona treatment to change the fiber surface physically. This helps to create rough surfaces that are beneficial for mechanical interlocking [122–124]. However, plasma treatment can be used to introduce chemical functionality onto the fibers so that they can undergo chemical reaction with the polymer matrix [122–124]. Surface treatment introduces rigidity and stiffness to the fibers and by extension, the composites will be suitable for possible structural applications in industries such as building, aerospace marine, packaging, automobile, etc.

Liu et al. [125] reported the improved mechanical characteristics, resulting from alkali treatment on Indian grass fiber used as reinforcement. The tensile Young's modulus of grass fiber reinforced composites treated with 30 wt% alkali improved by 60%, while their impact and flexural strengths improved by 30 and 40% respectively when compared to the 30 wt% raw fiber refeinforced composites. The structural and morphological changes of alkali modified kenaf fibers modified using: 3, 6 and 9% NaOH solutions were reported by Edeerozey et al. [126]. From their SEM results, it was found that 6% and 9% NaOH treatment gave better results. However, the 9% NaOH treatment was considered to be too strong for the fibers, as it led to low tensile properties.

5.2 Blending

Polymer blends have been defined as the combination of two or more polymers, prepared to enhance the properties of the products as well as to lessen the cost. Blends of polymer can simply be viewed as a polymer alloy. Therefore, blending is the mixing of two or more distinct polymers, in a way that gives a new material which has properties that are slightly different from the singular polymers involved. Polymer blending is a versatile procedure to obtain new polymer materials with characteristics far better than the individual polymers [127, 128]. It is a well-recognized option to solving emerging problem related to application requirements.

The ability to combine existing polymers into new one with superior qualities that are commercially acceptable makes blending a better option [129]. There are large numbers of polymer blends reported in literature by researchers in academics and industry [130–132]. Polymer blending can give rise to miscible or phase separated blends. For miscible blends, mixing approaches a molecular dimensional scale and the properties of the new polymer blend are more like those of a single polymer. While phase separated blends still exhibit the different properties of the polymers involved [133–136]. Although, miscibility is extremely rare in blends, the advantage of miscible blends over phase separated blends is in the property profile, especially mechanical properties. Most phase separated blends exhibit inconsistency in their mechanical properties due to the poor adhesion at the interface of the blend phases. However, researchers have developed specific methods by which these phase separation problems can be alleviated [129]. Hence, polymer matrix can be designed to suit the required properties expected of the composites. Blending also helps to modify the matrix with specific chemical functionality that could be added deliberately to enable good interaction between the fibers and the blended matrix. Many researchers have shown that the performance properties of blends can be heightened significantly by reinforcing them with either synthetic or natural fibers [137–141]. However, bio-fiber reinforcements have gained ground in recent time because of their numerous advantages over glass and carbon fibers [142].

According to Linares et al. [143], polymer blends have a recognized potential to produce high performance materials, however the polymer combination must be carefully selected. Müller-Buschbaum et al. [144] showed that blend composition has great influence on surface topography which is just one of the many factors that may affect the properties of the blend. Therefore, in the preparation of any blend, the selection of complementary polymers that will give the right kind of material hybrid with the required properties is the most important step. For a biodegradable polymer that will be green in all its ramifications, a blend of biopolymer material with a biodegradable counterpart is highly desirable. Among these, are: polycaprolactone, polybutylene succinate, etc. The use of blend started some decades ago [145], however, in the present day, the understanding of miscibility has undergone several changes [146]. The broad range of application of polymer materials requires varying properties according to the specific application area. With homopolymers, the range usually calls for special surface treatment or in some cases new polymer synthesis for each and every application. However, an economical alternative could be the preparing blends from specifically designated homopolymers [144].

5.3 Compatibilization

Compatibilization is described as the addition of a chemical substance to an immiscible or phase separated blend that help increase their stability. Compatibilizers are also referred to as coupling agents. They react at the interface of the blend to stabilize the phases. That is, they help to improve the compatibility between the two phases and increase miscibility [146]. The high interfacial tension caused by coalescence phase separated blends can be reduced by the compatibilizers, allowing a continuous flow of externally applied stress from the matrix to the fibers [130]. Compatibilization can be done by (i) introduction of specific interacting groups, (ii) in-situ polymerization grafting, (iii) addition of a ternary polymeric component, (iv) addition of block co-polymers of the blend polymers, (v) interpenetrating networks of crosslinked system and (vi) using reactive compatibilization methods [147–150]. For instance, polyhydoxyether of bisphenol A (Phenoxy: PHE) a non-reactive compatibilizer has been noted in some reports to have provided improved interfacial adhesion between immiscible and marginally compatible polymer blends [129]. Specifically, the addition of PHE to some polymers blends led to the blends yielding improved dispersion of the polymers within the blend, gave uniform injection molded surfaces and considerably increased the notched impact strength of the blends (polysulfone (PDF)/ABS; PSF/PA, PMMA/PA6 and PHE/PBT) [129, 151, 152]. Also reactive compatibilizers can be design to react with the fiber when introduced during processing. This method will involve modifying the fiber surfaces prior to the composite preparation. If successful, this method has the capability to improve interfacial interaction and subsequently improve mechanical properties.

5.4 Addition of nanoparticles

The use of nanoparticles to enhance the properties of NFPCs have been widely reported [99, 153, 154]. Nanoparticles are inorganic materials which possess a very high surface to volume ratio in which one or its entire dimension is less than 100 nm [155]. The addition of these nanoparticles influences the crystallization process during the solidification of the polymer composites leading to improved mechanical and thermal properties. These particles can be modified to selectively interact with a particular phase of the composites in a controlled manner [23, 156–158]. They can also be modified to act as a compatibilizer and react with both the matrix and the fibers to bring about good interaction and a better composite with stress transfer behavior.

Vargas et al. [159] reported the influence of nanofilters on some properties of polypropylene including mechanical properties. Their findings revealed that the nanofilters, in the presence of PP grafted MA, improved the tensile strength and Young's modulus properties which are indicative of the synergistic effect between the nanoparticles and compatibilizers. According to Lee and Youn [160], the addition of layered silicates worsened tensile properties of PP nanocomposites prepared by them. Similar investigation was presented by Rault et al. revealing that the addition of a maximum 1 wt% led to improvement in tensile properties but above the maximum, the silicate nanoparticle caused difficulties for processing the composites due to the formation of aggregates [161]. However, Joshi et al. [162] has reported improvement in tensile properties of PP/nanoclay composites. Therefore, we can conclude with certainty that there are other factors interfering in the positive influence expected from the addition of nanoparticle such as clay. According the Vargas et al. [159], fibers geometry plays a vital role in determining the composites eventual properties. Nanoparticles of different shapes and sizes were used to prepare polyamide 6 (PA6) composites in some research work conducted by Vlasveld et al. [163]. Their findings revealed that the rheological properties of the composites samples were highly dependent on the aspect ratio of the nanoparticle used. Therefore utmost care must be taken when the option of nanoparticles are being considered. Other nanoparticles have been used and their influence on the mechanical properties have been positive and very encouraging [164, 165].

5.5 Hybridization

Hybrid composites involve the combination of two or more different types, shapes or sizes of reinforcement in one composite material [166]. The hybrid composite properties have been reported to depend on many factors such as; fibers individual property, fiber-matrix compatibility, roughness of fiber surfaces, orientation of fibers and the extent of their intermingling [167]. Recently, investigations on the hybrid composites properties were based on the natural/synthetic fibers, natural/natural fiber and natural/synthetic/additive modified reinforced polymer composites. The popularity of these types of composites are increasing rapidly owning to their capability to provide freedom to tailor the composites and realizing properties that cannot be obtained in fabricated composites containing singular type reinforcement [21, 79].

Fiber-based hybrid composites have been reported to have improved properties compared with the unhybridized composites having single reinforcement [168–170]. Many research reports have shown that the addition of synthetic fibers at various amount to form hybrid fibers have composites of better-quality, especially in respect to their mechanical properties [171–179]. According to Ashik and Sharma [180], in one of their reviews, they listed some factors that may impact the mechanical properties of natural fiber hybrid polymeric composites, with the processing parameters featuring as one of the factors. Also, Nunna et al. [181] listed fiber content, fiber treatment, and the environmental conditions as some of the conditions that affect the properties of hybrid composites. For hybrid natural/synthetic fibers, it has been reported that as wt% of the synthetic fiber rises, the mechanical properties also rises. However, at a certain wt% of the synthetic fiber content added, the properties of the composites mechanical properties starts dropping and this may be ascribed to poor interfacial adhesion, high fiber-to-fiber contact, and poor wettability.

Mishra et al. [182] prepared hybrid glass fiber (GF) and pineapple leaf fiber (PLF) polyester composite with a total fiber content of 25 wt%. The tensile strength was observed to have increased as the GF was increased from 0 to 7.5 wt% to approximately 70 MPa, after which the strength started dropping. The flexural strength kept increasing as glass content increased from 4.3 to 12.9 wt%, in the hybrid biocomposite. The authors also prepared a similar hybrid biocomposite, but this time with 30 wt% total fibers (sisal and glass fibers). They reportedly observed a major improvement in tensile strength as the GF content was increased to 5.7 wt%. After this, the tensile strength was almost static even as the GF content was increased above this value. A similar trend was observed for the flexural strength. However, comparing the hybrid biocomposites with the biocomposites containing only pineapple or sisal fibers, the hybrid showed better improvement. Nevertheless, more evidence is available to show that the overall properties of hybrid composites depend greatly on (i) the percentage elongation at break and (ii) Young's modulus of the reinforcing fibers present.

Shahzad [183] presented the impact and fatigue properties of hybrid biocomposites of hemp and chopped strand mat glass fibers using unsaturated polyester resin as the matrix. Two different variations of hybrids composites were prepared. The first denoted with "A" containing 35.8 wt% hemp fiber and 11.1 wt% GF, while the second denoted with "B" had 36.6 wt% and 11.3 wt% hemp and glass fibers respectively. From the results, "A" had 70.1 ± 10.2 MPa, 8.3 ± 0.4 GPa and 1.31 ± 0.25% for tensile strength, Young's modulus and strain to failure respectively, while "B" showed 81.6 ± 3.7 MPa, 7.7 ± 0.3 GPa and 1.73 ± 0.08% for tensile strength, Young's modulus and strain to failure improvement respectively when compared with biocomposites reinforced only with hemp fibers, having

46.4 ± 4.6 MPa, 7.2 ± 0.9 and 1.03% respectively. The increase in the percentage of GF led to the observed increase in tensile strength and better strain to failure in "B" as compare to "A." The hemp fiber is a low strain to failure fiber while the glass fiber is a high strain to failure fiber. Their combination, leads to enhance strain to failure composites. This is referred to as "hybrid effect" and it has being well observed in hybrid composites. Therefore the increase in strain to failure of the hemp-glass fiber composites can be attributed to the hybrid effect. Also, there was an observed improvement in the fatigue strength of hybrid biocomposites while the fatigue sensitivity showed no improvements when compared to hemp only fiber composites.

Hybrid of glass/natural fibers have been reported to have improved impact, tensile and flexural strength [182]. Furthermore, Velmurugan and Manikandan [184] reported that good strength, especially mechanical strength is achieved when the synthetic fiber is placed at the ends of the composite for laminated composites.

In another research work, Ahmed and Vijayarangan [185] prepared composites with jute only and jute/glass fibers hybrid reinforced polyester composites, keeping the total weight fraction of fibers constant at 42 wt%. From their results, the composites consisting of 40:60 ratio of jute:glass fibers, the reinforced hybrid laminate gave an increase in the tensile strength, Young's modulus, flexural strength and flexural Young's modulus of 53, 30, 31 and 62% respectively over those of the jute only fiber composites. They further indicated that in the event properties, environmental impact and costs were to be considered, composites with 60:40 fiber fraction of jute:glass fibers ratio gave optimum material combination. This clearly highlights that the type of matrix and/or fiber, method of preparation, fiber content and fiber modification have a huge impact on the mechanical properties of hybrid biocomposites.

5.6 Other factors

In the preparation of NFPCs, there are many other additives or processes that can also influence the final properties of the composites but are seldom considered, for example impurities in the polymer matrix are introduced via fibers addition. These elements can influence the process of crystallization, just like the nanoparticles, although, this depends greatly on their chemical nature. Also it is worthy to mention that the annealing of the composites allows the crystals to grow to their maximum size. Thus, the temperature at which the materials are annealed is also very influential to the final properties of the composite obtained. Therefore from the selection of materials to the final product, care must be taken to achieve the properties desired.

Liu et al. [186] assessed the influence of processing method on the physical properties, especially mechanical properties, of kenaf fiber reinforced biocomposites prepared using soy fiber. The compression molded specimens were observed to have similar Young's modulus to those from injection molding at room temperature. However, at elevated temperature, the heat deflection temperature (HDT) and notched Izod impact strength were higher compared to those obtained from injection molded specimen. The improvements observed with the compression molded samples were attributed to a surge in Young's modulus at high temperature and fiber bridging effects.

Generally, biomedical applications desire for fabrication of grafts which are biocompatible and enable cell differentiation and expression with apt mechanical properties, but the ability to achieve such mechanical capacity has been a challenge for decades. As over the years, the focus and efforts have been geared towards biocompatibility and not necessarily mechanical prowess. But the formation of stresses occurring in implant locations due to mechanically inept implant materials have

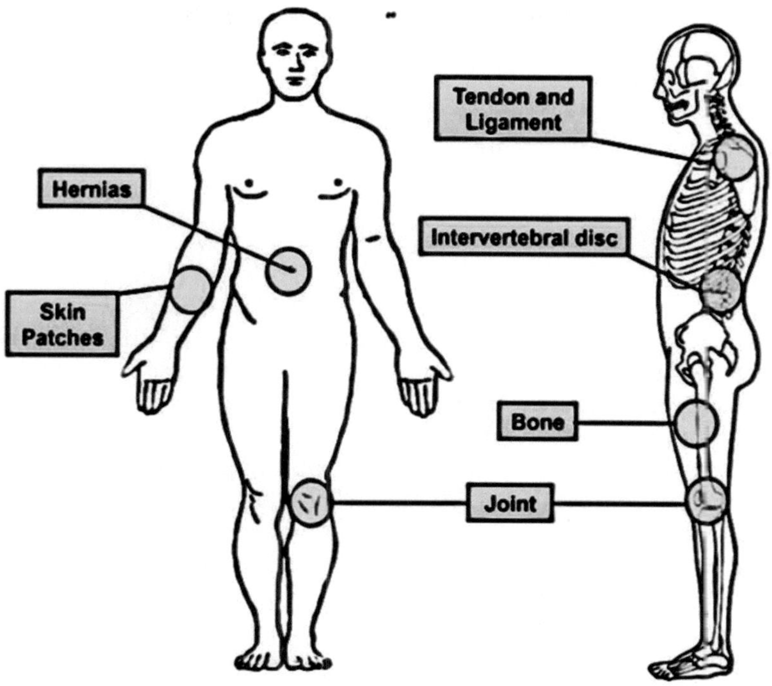

Figure 2.
Illustration of potential application of collagen fibers embedded in a hydrogel bio-composite for medical applications with adjusted mechanical properties that provide support and allow motion and flexibility of the tissue under repair [187].

No.	Patent title	Patent no.	Year
1.	Medical balloon with incorporated fibers	W0/2013/148399	2013
2.	Polyester cool-fiber antibacterial pillow	CN102715804	2012
3.	Medical natural porous fiber filler and vacuum sealing drainage device thereof	CN102715983	2012
4.	Manufacturing process of antibacterial bamboo pulp used for high-wet-modulus fiber	CN102677504	2012
5.	Flushable moist wipe or hygiene tissue	CN102665510-	2012
6.	Far-infrared fiber fabric functional bellyband by utilizing nano-selenium, germanium and zinc elements traditional Chinese medicine	CN10170[illegible]317	2010
7.	Medical device for insertion into a joint	US200902[illegible]4459	2008
8.	Medical device for insertion into a joint	EU1896[illegible]88	2008
9.	Antiviral fiber and producing method and use thereof	CN1609[illegible]36	2006
10.	Manufacturing of nano-fibers, from natural fibers, agro based fibers and root fibers	CA2437616	2005
11.	Natural antibacterial material and its use	CN461827	2003
12.	Absorbable protective coatings for wound with the use of sponge and process for producing the same	W0/20021054998	2006
13.	Medical prosthesis, especially for aneurysms, with a connection between its liner and its structure	EPOB18[illegible]84	1998

Table 1.
Summary of some patents published employing nature fibers for biomedical applications (adapted from Namvar et al. [188]).

led to numerous implant failures and lead to investigations to improve mechanical properties of biomedical devices for diverse applications (**Figure 2**).

There have been quite a number of patents registered over the years of researchers employing natural fiber for biomedical applications and **Table 1** presents a summary of some of these patents.

6. Conclusion

The natural fiber polymer composites have gained a lot of ground in terms of acceptance and applications. For these interests to increase continuously, the materials must be designed to meet certain requirements for their specific applicability. For structural applications, the most important property that is of concern is the mechanical properties as they help to predict the behavior of the materials under stress. As discussed in this chapter, NFPCs are being applied in the building and construction works, automobiles, aerospace, packaging, electronics and biomedical devices. The mechanical strength and toughness require by these various industries are quite different and can vary widely from one application to another. Even though the mechanical requirement varies, the factors that determine the mechanical properties are the same. These factors include type of fibers used, source of the fibers, surface treatment and modification carried out on the fibers, type of polymer matrix used (pristine and blended), type of fiber-matrix bond formed, and the internal arrangement which depends on the type of curing treatment after processing or annealing. In addition to this list are the fiber length, fiber orientation and distribution, fiber loading or volume faction, the type of functionality present and the extent of modification. All these factors can be manipulated to give a combination of the right measure of mechanical strength and stiffness or toughness required for the application it is being designed for. The use of compatibilizers and nanoparticles to modify the composites for specific purpose has been widely reported to improve the mechanical properties as well, but proper integrations must be considered. Furthermore, the use of hybrid fibers has also gained wide acceptance because of the improved stiffness and strength owing to the synergy observed in such fiber combination. The possibilities of NFPCs can be best imagined with the right improvement in their mechanical properties and this chapter has highlighted some of these benefits as presented by numerous research investigations across diverse fields.

Acknowledgements

The financial assistance of the University of Zululand and the National Research Foundation, South Africa through the South African Research Chair Initiative (SARChI) is hereby acknowledged. OSJ thanks the National Research Foundation (NRF) for a postdoctoral fellowship and funding under South African Research Chair for Nanotechnology.

Author details

Owonubi J. Shesan[1*†], Agwuncha C. Stephen[2†], Anusionwu G. Chioma[3], Revaprasadu Neerish[1] and Sadiku E. Rotimi[4,5]

1 Department of Chemistry, University of Zululand, Kwadlangezwa, KwaZulu-Natal, South Africa

2 Department of Chemistry, Ibrahim Badamasi Babangida University, Lapai, Nigeria

3 Department of Applied Chemistry, University of Johannesburg, Johannesburg, South Africa

4 Department of Chemical, Metallurgical and Material Engineering, Tshwane University of Technology, Pretoria, South Africa

5 Institute of Nano Engineering Research (INER), Tshwane University of Technology, Pretoria, South Africa

*Address all correspondence to: oshesan@gmail.com

† These authors contributed equally.

References

[1] Gao H, Wang J, Chen X, Wang G, Huang X, Li A, et al. Nanoconfinement effects on thermal properties of nanoporous shape-stabilized composite PCMs: A review. Nano Energy. 2018;**53**:769-797

[2] Kargarzadeh H, Mariano M, Huang J, Lin N, Ahmad I, Dufresne A, et al. Recent developments on nanocellulose reinforced polymer nanocomposites: A review. Polymer. 2017;**132**:368-393

[3] Farah S, Anderson DG, Langer R. Physical and mechanical properties of PLA, and their functions in widespread applications—A comprehensive review. Advanced Drug Delivery Reviews. 2016;**107**:367-392

[4] Kahl C, Feldmann M, Sälzer P, Heim H-P. Advanced short fiber composites with hybrid reinforcement and selective fiber-matrix-adhesion based on polypropylene–Characterization of mechanical properties and fiber orientation using high-resolution X-ray tomography. Composites Part A: Applied Science and Manufacturing. 2018;**111**:54-61

[5] Thakur VK, Thakur MK. Processing and characterization of natural cellulose fibers/thermoset polymer composites. Carbohydrate Polymers. 2014;**109**:102-117

[6] Deepa B, Abraham E, Cordeiro N, Mozetic M, Mathew AP, Oksman K, et al. Utilization of various lignocellulosic biomass for the production of nanocellulose: A comparative study. Cellulose. 2015;**22**(2):1075-1090

[7] Lee H-R, Kim K, Mun SC, Chang YK, Choi SQ. A new method to produce cellulose nanofibrils from microalgae and the measurement of their mechanical strength. Carbohydrate Polymers. 2018;**180**:276-285

[8] da Luz FS, Ramos FJHTV, Nascimento LFC, da Silva Figueiredo AB-H, Monteiro SN. Critical length and interfacial strength of PALF and coir fiber incorporated in epoxy resin matrix. Journal of Materials Research and Technology. 2018;**7**(4):528-534

[9] Raju B, Hiremath S, Mahapatra DR. A review of micromechanics based models for effective elastic properties of reinforced polymer matrix composites. Composite Structures. 2018;**204**:607-619

[10] Sen T, Reddy HJ. Various industrial applications of hemp, kinaf, flax and ramie natural fibres. International Journal of Innovation, Management and Technology. 2011;**2**(3):192

[11] Nechyporchuk O, Belgacem MN, Bras J. Production of cellulose nanofibrils: A review of recent advances. Industrial Crops and Products. 2016;**93**:2-25

[12] Mittal G, Rhee KY, Mišković-Stanković V, Hui D. Reinforcements in multi-scale polymer composites: Processing, properties, and applications. Composites Part B: Engineering. 2018;**138**:122-139

[13] Spitalsky Z, Tasis D, Papagelis K, Galiotis C. Carbon nanotube–polymer composites: Chemistry, processing, mechanical and electrical properties. Progress in Polymer Science. 2010;**35**(3):357-401

[14] Naveen MH, Gurudatt NG, Shim Y-B. Applications of conducting polymer composites to electrochemical sensors: A review. Applied Materials Today. 2017;**9**:419-433

[15] Merlini C, Soldi V, Barra GM. Influence of fiber surface treatment and length on physico-chemical properties of short random banana

fiber-reinforced castor oil polyurethane composites. Polymer Testing. 2011;**30**(8):833-840

[16] Thoppul SD, Finegan J, Gibson RF. Mechanics of mechanically fastened joints in polymer–matrix composite structures—A review. Composites Science and Technology. 2009;**69**(3-4):301-329

[17] Thakur A, Purohit R, Rana R, Bandhu D. Characterization and evaluation of mechanical behavior of epoxy-CNT-bamboo matrix hybrid composites. Materials Today: Proceedings. 2018;**5**(2):3971-3980

[18] Zhao W, Li T, Li Y, O'Brien DJ, Terrones M, Wei B, et al. Mechanical properties of nanocomposites reinforced by carbon nanotube sponges. Journal of Materiomics. 2018;**4**:157-164

[19] Chung D. Processing-structure-property relationships of continuous carbon fiber polymer-matrix composites. Materials Science and Engineering: R: Reports. 2017;**113**:1-29

[20] Supian A, Sapuan S, Zuhri M, Syams Z, Ya HH. Hybrid reinforced thermoset polymer composite in energy absorption tube application: A review. Defence Technology. 2018;**14**:291-305

[21] Väisänen T, Das O, Tomppo L. A review on new bio-based constituents for natural fiber-polymer composites. Journal of Cleaner Production. 2017;**149**:582-596

[22] Ferreira F, Dufresne A, Pinheiro I, Souza D, Gouveia R, Mei L, et al. How do cellulose nanocrystals affect the overall properties of biodegradable polymer nanocomposites: A comprehensive review. European Polymer Journal. 2018;**108**:274-285

[23] Zhang X, Xu Y, Zhang X, Wu H, Shen J, Chen R, et al. Progress on the layer-by-layer assembly of multilayered polymer composites: Strategy, structural control and applications. Progress in Polymer Science. 2018;**89**:76-107

[24] Del Nobile MA, Conte A, Buonocore GG, Incoronato A, Massaro A, Panza O. Active packaging by extrusion processing of recyclable and biodegradable polymers. Journal of Food Engineering. 2009;**93**(1):1-6

[25] Seabra AB, Bernardes JS, Fávaro WJ, Paula AJ, Durán N. Cellulose nanocrystals as carriers in medicine and their toxicities: A review. Carbohydrate Polymers. 2018;**181**:514-527

[26] Feng Y-H, Cheng T-Y, Yang W-G, Ma P-T, He H-Z, Yin X-C, et al. Characteristics and environmentally friendly extraction of cellulose nanofibrils from sugarcane bagasse. Industrial Crops and Products. 2018;**111**:285-291

[27] Arrieta MP, Fortunati E, Dominici F, López J, Kenny JM. Bionanocomposite films based on plasticized PLA–PHB/cellulose nanocrystal blends. Carbohydrate Polymers. 2015;**121**:265-275

[28] Bourmaud A, Beaugrand J, Shah DU, Placet V, Baley C. Towards the design of high-performance plant fibre composites: How can we best define the diversity and specificities of plant cell walls? Progress in Materials Science. 2018;**97**:347-408

[29] Chávez-Guerrero L, Sepúlveda-Guzmán S, Silva-Mendoza J, Aguilar-Flores C, Pérez-Camacho O. Eco-friendly isolation of cellulose nanoplatelets through oxidation under mild conditions. Carbohydrate Polymers. 2018;**181**:642-649

[30] Ayre D. Technology advancing polymers and polymer composites towards sustainability: A review. Current Opinion in Green and Sustainable Chemistry. 2018;**13**:108-112

[31] Brand J, Pecastaings G, Sèbe G. A versatile method for the surface tailoring of cellulose nanocrystal building blocks by acylation with functional vinyl esters. Carbohydrate Polymers. 2017;**169**:189-197

[32] Campo EA. Selection of Polymeric Materials: How to Select Design Properties from Different Standards. Norwich, NY: William Andrew; 2008

[33] Zweben C. Tensile strength of hybrid composites. Journal of Materials Science. 1977;**12**(7):1325-1337

[34] Fernandes F, Tavares J, de Sousa RA, Pereira A, Esteves J. Manufacturing and testing composites based on natural materials. Procedia Manufacturing. 2017;**13**:227-234

[35] Ptak M, Kaczyński P, Wilhelm J, Margarido J, Marques P, Pinto S, et al. Graphene-enriched agglomerated cork material and its behaviour under quasi-static and dynamic loading. Materials. 2019;**12**(1):151

[36] Swolfs Y, Gorbatikh L, Verpoest I. Fibre hybridisation in polymer composites: A review. Composites Part A: Applied Science and Manufacturing. 2014;**67**:181-200

[37] García A, Gandini A, Labidi J, Belgacem N, Bras J. Industrial and crop wastes: A new source for nanocellulose biorefinery. Industrial Crops and Products. 2016;**93**:26-38

[38] Chen J, Yan L, Song W, Xu D. Interfacial characteristics of carbon nanotube-polymer composites: A review. Composites Part A: Applied Science and Manufacturing. 2018;**114**:149-169

[39] Benkaddour A, Jradi K, Robert S, Daneault C. Study of the effect of grafting method on surface polarity of tempo-oxidized nanocellulose using polycaprolactone as the modifying compound: Esterification versus click-chemistry. Nanomaterials. 2013;**3**(4):638-654

[40] Osman B, Özer ET, Beşirli N, Güçer Ş. Development and application of a solid phase extraction method for the determination of phthalates in artificial saliva using new synthesised microspheres. Polymer Testing. 2013;**32**(4):810-818

[41] Boufi S, Belgacem MN. Modified cellulose fibres for adsorption of dissolved organic solutes. Cellulose. 2006;**13**(1):81-94

[42] Bas O, De-Juan-Pardo EM, Catelas I, Hutmacher DW. The quest for mechanically and biologically functional soft biomaterials via soft network composites. Advanced Drug Delivery Reviews. 2018;**132**:214-234

[43] Benyahia H, Tarfaoui M, El Moumen A, Ouinas D, Hassoon O. Mechanical properties of offshoring polymer composite pipes at various temperatures. Composites Part B: Engineering. 2018;**152**:231-240

[44] De Greef N, Gorbatikh L, Godara A, Mezzo L, Lomov SV, Verpoest I. The effect of carbon nanotubes on the damage development in carbon fiber/epoxy composites. Carbon. 2011;**49**(14):4650-4664

[45] De Greef N, Gorbatikh L, Lomov SV, Verpoest I. Damage development in woven carbon fiber/epoxy composites modified with carbon nanotubes under tension in the bias direction. Composites Part A: Applied Science and Manufacturing. 2011;**42**(11):1635-1644

[46] Qian H, Greenhalgh ES, Shaffer MS, Bismarck A. Carbon nanotube-based hierarchical composites: A review. Journal of Materials Chemistry. 2010;**20**(23):4751-4762

[47] Dweib M, Hu B, Shenton H III, Wool R. Bio-based composite roof

structure: Manufacturing and processing issues. Composite Structures. 2006;**74**(4):379-388

[48] Bektas I, Guler C, Kalaycioğlu H, Mengeloglu F, Nacar M. The manufacture of particleboards using sunflower stalks (*Helianthus annuus* L.) and poplar wood (*Populus alba* L.). Journal of Composite Materials. 2005;**39**(5):467-473

[49] John MJ, Thomas S. Biofibres and biocomposites. Carbohydrate Polymers. 2008;**71**(3):343-364

[50] Li Q, Matuana LM. Surface of cellulosic materials modified with functionalized polyethylene coupling agents. Journal of Applied Polymer Science. 2003;**88**(2):278-286

[51] Owonubi SJ, Linganiso LZ, Motaung TE, Songca SP. Wood and its byproducts: Cascading utilisation for biomass (re)generation. In: Linganiso LZ, Motaung TE, editors. "Waste-to-Profit" (W-t-P): Circular Economy in the Construction Industry for a Sustainable Future. Unites States of America: Nova Publishers; 2018

[52] Coutts RS. A review of Australian research into natural fibre cement composites. Cement and Concrete Composites. 2005;**27**(5):518-526

[53] Peijs T. Composites for recyclability. Materials Today. 2003;**4**(6):30-35

[54] Dunne R, Desai D, Sadiku R, Jayaramudu J. A review of natural fibres, their sustainability and automotive applications. Journal of Reinforced Plastics and Composites. 2016;**35**(13):1041-1050

[55] Bos HL. The potential of flax fibres as reinforcement for composite materials. Eindhoven: Technische Universiteit Eindhoven; 2004

[56] Brief L. Opportunities in Natural Fiber Composites. Irving (TX): Lucintel LLC; 2011

[57] La Mantia F, Morreale M. Green composites: A brief review. Composites Part A: Applied Science and Manufacturing. 2011;**42**(6):579-588

[58] Marsh G. Next step for automotive materials. Materials Today. 2003;**6**(4): 36-43

[59] La Mantia F, Morreale M. Mechanical properties of recycled polyethylene ecocomposites filled with natural organic fillers. Polymer Engineering & Science. 2006;**46**(9):1131-1139

[60] Netravali AN, Chabba S. Composites get greener. Materials Today. 2003;**4**(6):22-29

[61] Kiruthika A. A review on physico-mechanical properties of bast fibre reinforced polymer composites. Journal of Building Engineering. 2017;**9**:91-99

[62] Plackett D, Andersen TL, Pedersen WB, Nielsen L. Biodegradable composites based on L-polylactide and jute fibres. Composites Science and Technology. 2003;**63**(9):1287-1296

[63] Li Y, Mai Y-W, Ye L. Sisal fibre and its composites: A review of recent developments. Composites Science and Technology. 2000;**60**(11):2037-2055

[64] Senthilkumar K, Saba N, Rajini N, Chandrasekar M, Jawaid M, Siengchin S, et al. Mechanical properties evaluation of sisal fibre reinforced polymer composites: A review. Construction and Building Materials. 2018;**174**:713-729

[65] Pickering KL, Efendy MA, Le TM. A review of recent developments in natural fibre composites and their mechanical performance. Composites Part A: Applied Science and Manufacturing. 2016;**83**:98-112

[66] Pickering KL, Beckermann G, Alam S, Foreman NJ. Optimising

industrial hemp fibre for composites. Composites Part A: Applied Science and Manufacturing. 2007;**38**(2):461-468

[67] Saba N, Paridah M, Jawaid M. Mechanical properties of kenaf fibre reinforced polymer composite: A review. Construction and Building Materials. 2015;**76**:87-96

[68] Kabir M, Wang H, Lau K, Cardona F. Chemical treatments on plant-based natural fibre reinforced polymer composites: An overview. Composites Part B: Engineering. 2012;**43**(7):2883-2892

[69] Kabir M, Wang H, Aravinthan T, Cardona F, Lau K-T. Effects of natural fibre surface on composite properties: A review. In: Proceedings of the 1st International Postgraduate Conference on Engineering, Designing and Developing the Built Environment for Sustainable Wellbeing (eddBE2011). Australia: Queensland University of Technology; 2011

[70] Phillips DM, Pierce MR, Baur JW. Mechanical and thermal analysis of microvascular networks in structural composite panels. Composites Part A: Applied Science and Manufacturing. 2011;**42**(11):1609-1619

[71] Megiatto JD Jr, Silva CG, Ramires EC, Frollini E. Thermoset matrix reinforced with sisal fibers: Effect of the cure cycle on the properties of the biobased composite. Polymer Testing. 2009;**28**(8):793-800

[72] Fakirov S, Bhattacharyya D. Handbook of Engineering Biopolymers: Homopolymers, Blends, and Composites. Germany: Carl Hanser Verlag GmbH Co KG; 2007

[73] Joseph P. Studies on Short Sisal Fibre Reinforced Isotactic Polypropylene Composites. 2001. Available from: http://shodhganga.inflibnet.ac.in:8080/jspui/handle/10603/6193?mode=full

[74] Ku H, Wang H, Pattarachaiyakoop N, Trada M. A review on the tensile properties of natural fiber reinforced polymer composites. Composites Part B: Engineering. 2011;**42**(4):856-873

[75] Ahmad I, Baharum A, Abdullah I. Effect of extrusion rate and fiber loading on mechanical properties of Twaron fiber-thermoplastic natural rubber (TPNR) composites. Journal of Reinforced Plastics and Composites. 2006;**25**(9):957-965

[76] Kargarzadeh H, Huang J, Lin N, Ahmad I, Mariano M, Dufresne A, et al. Recent developments in nanocellulose-based biodegradable polymers, thermoplastic polymers, and porous nanocomposites. Progress in Polymer Science. 2018;**87**:197-227

[77] Ji X, Xu Y, Zhang W, Cui L, Liu J. Review of functionalization, structure and properties of graphene/polymer composite fibers. Composites Part A: Applied Science and Manufacturing. 2016;**87**:29-45

[78] Inácio AL, Nonato RC, Bonse BC. Mechanical and thermal behavior of aged composites of recycled PP/EPDM/talc reinforced with bamboo fiber. Polymer Testing. 2018;**72**:357-363

[79] Lila MK, Singhal A, Banwait SS, Singh I. A recyclability study of bagasse fiber reinforced polypropylene composites. Polymer Degradation and Stability. 2018;**152**:272-279

[80] Wang B, Panigrahi S, Tabil L, Crerar W. Pre-treatment of flax fibers for use in rotationally molded biocomposites. Journal of Reinforced Plastics and Composites. 2007;**26**(5):447-463

[81] Alamri H, Low IM. Mechanical properties and water absorption behaviour of recycled cellulose fibre reinforced epoxy composites. Polymer Testing. 2012;**31**(5):620-628

[82] Dhal JP, Mishra SC. Processing and properties of natural fiber-reinforced polymer composite. Journal of Materials. 2012;**2013**:6. Article ID: 297213

[83] Vroman I, Tighzert L. Biodegradable polymers. Materials. 2009;**2**(2):307-344

[84] Mallick PK. Fiber-reinforced Composites: Materials, Manufacturing, and Design. Boca Raton, Florida, United States: CRC Press; 2007

[85] Angelov I, Wiedmer S, Evstatiev M, Friedrich K, Mennig G. Pultrusion of a flax/polypropylene yarn. Composites Part A: Applied Science and Manufacturing. 2007;**38**(5):1431-1438

[86] Rodríguez E, Petrucci R, Puglia D, Kenny JM, Vazquez A. Characterization of composites based on natural and glass fibers obtained by vacuum infusion. Journal of Composite Materials. 2005;**39**(3):265-282

[87] Tungjitpornkull S, Sombatsompop N. Processing technique and fiber orientation angle affecting the mechanical properties of E-glass fiber reinforced wood/PVC composites. Journal of Materials Processing Technology. 2009;**209**(6):3079-3088

[88] Siaotong B, Tabil L, Panigrahi S, Crerar W. Determination of optimum extrusion parameters in compounding flax fiber-reinforced polyethylene composites. In 2006 ASAE annual meeting. Michigan: American Society of Agricultural and Biological Engineers; 2006

[89] Li X, Panigrahi S, Tabil L. A study on flax fiber-reinforced polyethylene biocomposites. Applied Engineering in Agriculture. 2009;**25**(4):525-531

[90] Sreekumar P, Joseph K, Unnikrishnan G, Thomas S. A comparative study on mechanical properties of sisal-leaf fibre-reinforced polyester composites prepared by resin transfer and compression moulding techniques. Composites Science and Technology. 2007;**67**(3-4):453-461

[91] Hussain B, Safiulla M. Comparative study of cooling systems for vacuum forming tool. Materials Today: Proceedings. 2018;**5**(1):30-36

[92] Hammami A, Gebart B. Analysis of the vacuum infusion molding process. Polymer Composites. 2000;**21**(1):28-40

[93] Gu Y, Tan X, Yang Z, Zhang Z. Hot compaction and mechanical properties of ramie fabric/epoxy composite fabricated using vacuum assisted resin infusion molding. Materials & Design (1980-2015). 2014;**56**:852-861

[94] Vaxman A, Narkis M, Siegmann A, Kenig S. Void formation in short-fiber thermoplastic composites. Polymer Composites. 1989;**10**(6):449-453

[95] Mwaikambo LY, Ansell MP. Chemical modification of hemp, sisal, jute, and kapok fibers by alkalization. Journal of Applied Polymer Science. 2002;**84**(12):2222-2234

[96] Xie W, Zhu W, Shen Z. Synthesis, isothermal crystallization and micellization of mPEG–PCL diblock copolymers catalyzed by yttrium complex. Polymer. 2007;**48**(23):6791-6798

[97] Kalia S, Thakur K, Celli A, Kiechel MA, Schauer CL. Surface modification of plant fibers using environment friendly methods for their application in polymer composites, textile industry and antimicrobial activities: A review. Journal of Environmental Chemical Engineering. 2013;**1**(3):97-112

[98] Jain D, Vats S, Bera TK. Micromechanical interactions and their relation to stress field

for different classes of reinforced polymer composites. Materials Today: Proceedings. 2018;**5**(9):19944-19953

[99] Ibrahim ID, Jamiru T, Sadiku RE, Kupolati WK, Agwuncha SC. Dependency of the mechanical properties of sisal fiber reinforced recycled polypropylene composites on fiber surface treatment, fiber content and nanoclay. Journal of Polymers and the Environment. 2017;**25**(2):427-434

[100] Ibrahim ID, Jamiru T, Sadiku ER, Kupolati WK, Agwuncha SC. Impact of surface modification and nanoparticle on sisal fiber reinforced polypropylene nanocomposites. Journal of Nanotechnology. 2016:9. Article ID: 4235975

[101] Khoathane MC, Sadiku ER, Agwuncha CS. Surface ModIfIcatIon of natural fIber coMpoSIteS and theIr potentIal applIcatIonS. In: Surface Modification of Biopolymers. Hoboken, NJ, USA: John Wiley & Sons, Inc.; 2015. pp. 370-400

[102] Shekar HS, Ramachandra M. Green composites: A review. Materials Today: Proceedings. 2018;**5**(1):2518-2526

[103] Li Y, Hu C, Yu Y. Interfacial studies of sisal fiber reinforced high density polyethylene (HDPE) composites. Composites Part A: Applied Science and Manufacturing. 2008;**39**(4):570-578

[104] Sanjay M, Madhu P, Jawaid M, Senthamaraikannan P, Senthil S, Pradeep S. Characterization and properties of natural fiber polymer composites: A comprehensive review. Journal of Cleaner Production. 2018;**172**:566-581

[105] Alavi-Soltani S, Sabzevari S, Koushyar H, Minaie B. Thermal, rheological, and mechanical properties of a polymer composite cured at different isothermal cure temperatures. Journal of Composite Materials. 2012;**46**(5):575-587

[106] Liu X, Dai G. Surface modification and micromechanical properties of jute fiber mat reinforced polypropylene composites. Express Polymer Letters. 2007;**1**(5):299-307

[107] Tao Y, Yan L, Jie R. Preparation and properties of short natural fiber reinforced poly (lactic acid) composites. Transactions of Nonferrous Metals Society of China. 2009;**19**:s651-s655

[108] Yuan X, Zhu B, Cai X, Qiao K, Zhao S, Yu J. Influence of different surface treatments on the interfacial adhesion of graphene oxide/carbon fiber/epoxy composites. Applied Surface Science. 2018;**458**:996-1005

[109] Doan T-T-L, Gao S-L, Mäder E. Jute/polypropylene composites I. Effect of matrix modification. Composites Science and Technology. 2006;**66**(7-8):952-963

[110] Mondal S. Preparation, properties and applications of nanocellulosic materials. Carbohydrate Polymers. 2017;**163**:301-316

[111] Fortunati E, Luzi F, Yang W, Kenny JM, Torre L, Puglia D. Bio-Based Nanocomposites in Food Packaging, in Nanomaterials for Food Packaging. United Kingdom: Elsevier; 2018. pp. 71-110

[112] Fortunati E, Peltzer M, Armentano I, Jiménez A, Kenny JM. Combined effects of cellulose nanocrystals and silver nanoparticles on the barrier and migration properties of PLA nano-biocomposites. Journal of Food Engineering. 2013;**118**(1):117-124

[113] Alila S, Besbes I, Vilar MR, Mutjé P, Boufi S. Non-woody plants as raw materials for production of microfibrillated cellulose (MFC): A comparative study. Industrial Crops and Products. 2013;**41**:250-259

[114] El Achaby M, Kassab Z, Barakat A, Aboulkas A. Alfa fibers as viable

sustainable source for cellulose nanocrystals extraction: Application for improving the tensile properties of biopolymer nanocomposite films. Industrial Crops and Products. 2018;**112**:499-510

[115] Jonas R, Farah LF. Production and application of microbial cellulose. Polymer Degradation and Stability. 1998;**59**(1-3):101-106

[116] Brown AJ. XLIII. On an acetic ferment which forms cellulose. Journal of the Chemical Society, Transactions. 1886;**49**:432-439

[117] Delmer DP, Amor Y. Cellulose biosynthesis. The Plant Cell. 1995;**7**(7):987

[118] Shoda M, Sugano Y. Recent advances in bacterial cellulose production. Biotechnology and Bioprocess Engineering. 2005;**10**(1):1

[119] Panesar P, Chavan Y, Bera M, Chand O, Kumar H. Evaluation of Acetobacter strain for the production of microbial cellulose. Asian Journal of Chemistry. 2009;**21**(10):99-102

[120] Embuscado ME, Marks JS, BeMiller JN. Bacterial cellulose. I. Factors affecting the production of cellulose by *Acetobacter xylinum*. Food Hydrocolloids. 1994;**8**(5):407-418

[121] Steinbüchel, A. and S.K. Rhee, Polysaccharides and Polyamides in the Food Industry: Properties, Production, and Patents. Weinheim, Germany: Wiley-VCH Verlag GmbH & CO. KGaA; 2005

[122] Kale KH, Desai A. Atmospheric pressure plasma treatment of textiles using non-polymerising gases. Indian Journal of Fibre & Textile Research. 2011;**36**(3):289-299

[123] Kim B, Nguyen M, Hwang B, Lee S. Effect of plasma treatment on the mechanical properties of natural fiber/PP composites. WIT Transactions on the Built Environment. 2008;**97**:159-166

[124] Wolter M, Bornholdt S, Häckel M, Kersten H. Atmospheric pressure plasma jet for treatment of polymers. Journal of Achievements in Materials and Manufacturing Engineering. 2009;**37**(2):730-734

[125] Liu W, Mohanty AK, Askeland P, Drzal LT, Misra M. Influence of fiber surface treatment on properties of Indian grass fiber reinforced soy protein based biocomposites. Polymer. 2004;**45**(22):7589-7596

[126] Edeerozey AM, Akil HM, Azhar A, Ariffin MZ. Chemical modification of kenaf fibers. Materials Letters. 2007;**61**(10):2023-2025

[127] Stelescu DM, Airinei A, Homocianu M, Fifere N, Timpu D, Aflori M. Structural characteristics of some high density polyethylene/EPDM blends. Polymer Testing. 2013;**32**(2):187-196

[128] Farsi M. Thermoplastic matrix reinforced with natural fibers: A study on interfacial behavior. In: Some Critical Issues for Injection Molding. United Kingdom: INTECH; 2012

[129] Robeson LM. Polymer blends. A Comprehensive Review. In: Polymer Blends. München: Carl Hanser Verlag GmbH & Co. KG; 2007. pp. I-XII. Available from: https://www.hanser-elibrary.com/doi/pdf/10.3139/9783446436503.fm. eBook ISBN: 978-3-446-43650-3. Print ISBN: 978-3-446-22569-5 2007

[130] Yatigala NS, Bajwa DS, Bajwa SG. Compatibilization improves physico-mechanical properties of biodegradable biobased polymer composites. Composites Part A: Applied Science and Manufacturing. 2018;**107**:315-325

[131] Fox DM, Lee J, Citro CJ, Novy M. Flame retarded poly (lactic acid)

using POSS-modified cellulose. 1. Thermal and combustion properties of intumescing composites. Polymer Degradation and Stability. 2013;**98**(2):590-596

[132] Lin JH, Lu CT, He CH, Huang CC, Lou CW. Preparation and evaluation of artificial bone complex material: Chitosan/polylactic complex braids. Journal of Composite Materials. 2011;**45**(19):1945-1951

[133] Kemala T, Budianto E, Soegiyono B. Preparation and characterization of microspheres based on blend of poly (lactic acid) and poly (ε-caprolactone) with poly (vinyl alcohol) as emulsifier. Arabian Journal of Chemistry. 2012;**5**(1):103-108

[134] Choi NS, Kim CH, Cho KY, Park JK. Morphology and hydrolysis of PCL/PLLA blends compatibilized with P (LLA-co-ϵCL) or P (LLA-b-ϵCL). Journal of Applied Polymer Science. 2002;**86**(8):1892-1898

[135] Tuba F, Oláh L, Nagy P. Characterization of reactively compatibilized poly (d, l-lactide)/poly (ε-caprolactone) biodegradable blends by essential work of fracture method. Engineering Fracture Mechanics. 2011;**78**(17):3123-3133

[136] Younas M, Noreen A, Sharif A, Majeed A, Hssan A, Tabasum S, et al. A review on versatile applications of blends and composites of CNC with natural and synthetic polymers with mathematical modeling. International Journal of Biological Macromolecules. 2018;**124**:591-626

[137] Ho M-P, Wang H, Lau K-T, Lee J-H, Hui D. Interfacial bonding and degumming effects on silk fibre/polymer biocomposites. Composites Part B: Engineering. 2012;**43**(7):2801-2812

[138] Das D, Mukhopadhyay S, Kaur H. Optimization of fiber composition in natural fiber-reinforced composites using a simplex lattice design. Journal of Composite Materials. 2012;**46**(26):3311-3319

[139] Kaleemullah M, Khan SU, Kim J-K. Effect of surfactant treatment on thermal stability and mechanical properties of CNT/polybenzoxazine nanocomposites. Composites Science and Technology. 2012;**72**(16):1968-1976

[140] Fang K, Wang B, Sheng K, Sun XS. Properties and morphology of poly (lactic acid)/soy protein isolate blends. Journal of Applied Polymer Science. 2009;**114**(2):754-759

[141] Graupner N. Improvement of the mechanical properties of biodegradable hemp fiber reinforced poly (lactic acid)(PLA) composites by the admixture of man-made cellulose fibers. Journal of Composite Materials. 2009;**43**(6):689-702

[142] Wambua P, Ivens J, Verpoest I. Natural fibres: Can they replace glass in fibre reinforced plastics? Composites Science and Technology. 2003;**63**(9):1259-1264

[143] Linares EM, Rippel MM, Galembeck F. Clay platelet partition within polymer blend nanocomposite films by EFTEM. ACS Applied Materials & Interfaces. 2010;**2**(12):3648-3653

[144] Müller-Buschbaum P, Gutmann JS, Stamm M. Influence of blend composition on phase separation and dewetting of thin polymer blend films. Macromolecules. 2000;**33**(13):4886-4895

[145] White RP, Lipson JE, Higgins JS. New correlations in polymer blend miscibility. Macromolecules. 2012;**45**(2):1076-1084

[146] Utracki LA. Compatibilization of polymer blends. The Canadian

Journal of Chemical Engineering. 2002;**80**(6):1008-1016

[147] Yavas BH, Tanrıver N, Benli B, Kizilcan N. In situ polymerization of sepiolite modified polysulfone. Procedia-Social and Behavioral Sciences. 2015;**195**:2206-2209

[148] Cao E, Duan W, Wang F, Wang A, Zheng Y. Natural cellulose fiber derived hollow-tubular-oriented polydopamine: In-situ formation of Ag nanoparticles for reduction of 4-nitrophenol. Carbohydrate Polymers. 2017;**158**:44-50

[149] Rask M, Madsen B, Sørensen BF, Fife JL, Martyniuk K, Lauridsen EM. In situ observations of microscale damage evolution in unidirectional natural fibre composites. Composites Part A: Applied Science and Manufacturing. 2012;**43**(10):1639-1649

[150] Zapata P, Quijada R, Retuer J, Moncada E. Preparation of nanocomposites by in situ polimerization. Journal of the Chilean Chemical Society. 2008;**53**(1):1359-1360

[151] Remiro P, Nazabal J. Phase behavior of ternary PBT–PC/phenoxy blends. Journal of Applied Polymer Science. 1991;**42**(5):1475-1483

[152] Robeson LM, Claus Jr WD, Batleman HL. Poly (aryl ether) containing blends. Google Patents. 1983

[153] Agwuncha SC, Ray SS, Jayaramudu J, Khoathane C, Sadiku R. Influence of boehmite nanoparticle loading on the mechanical, thermal, and rheological properties of biodegradable polylactide/poly (ϵ-caprolactone) blends. Macromolecular Materials and Engineering. 2015;**300**(1):31-47

[154] Ibrahim ID, Jamiru T, Sadiku ER, Kupolati WK, Agwuncha SC, Ekundayo G. Mechanical properties of sisal fibre-reinforced polymer composites: A review. Composite Interfaces. 2016;**23**(1):15-36

[155] Owonubi SJ, Mukwevho E, Revaprasadu N. Nanoparticle-based delivery of plant metabolites. In: Goyal MR et al., editors. The Therapeutic Properties of Medicinal Plants: Health-Rejuvenating Bioactive Compounds of Native Flora. New York: CRC Press; 2019

[156] Bounor-Legaré V, Cassagnau P. In situ synthesis of organic–inorganic hybrids or nanocomposites from sol–gel chemistry in molten polymers. Progress in Polymer Science. 2014;**39**(8):1473-1497

[157] Eichner E, Heinrich S, Schneider G. Influence of particle shape and size on mechanical properties in copper-polymer composites. Powder Technology. 2018;**339**:39-45

[158] Huang X, Netravali AN. Characterization of nano-clay reinforced phytagel-modified soy protein concentrate resin. Biomacromolecules. 2006;7(10):2783-2789

[159] Vargas AF, Orozco VH, Rault F, Giraud S, Devaux E, López BL. Influence of fiber-like nanofillers on the rheological, mechanical, thermal and fire properties of polypropylene: An application to multifilament yarn. Composites Part A: Applied Science and Manufacturing. 2010;**41**(12):1797-1806

[160] Lee SH, Youn JR. Properties of polypropylene/layered-silicate nanocomposites and melt-spun fibers. Journal of Applied Polymer Science. 2008;**109**(2):1221-1231

[161] Rault F, Campagne C, Rochery M, Giraud S, Devaux E. Polypropylene multifilament yarn filled with clay and/or graphite: Study of a potential synergy. Journal of Polymer Science Part B: Polymer Physics. 2010;**48**(11):1185-1195

[162] Joshi SV, Drzal L, Mohanty A, Arora S. Are natural fiber composites environmentally superior to glass fiber reinforced composites? Composites Part A: Applied Science and Manufacturing. 2004;**35**(3):371-376

[163] Vlasveld D, Groenewold J, Bersee H, Picken S. Moisture absorption in polyamide-6 silicate nanocomposites and its influence on the mechanical properties. Polymer. 2005;**46**(26):12567-12576

[164] Özdilek C, Kazimierczak K, van der Beek D, Picken SJ. Preparation and properties of polyamide-6-boehmite nanocomposites. Polymer. 2004;**45**(15):5207-5214

[165] Kim D, Krishnamoorti R. Interfacial activity of poly [oligo (ethylene oxide)–monomethyl ether methacrylate]-grafted silica nanoparticles. Industrial & Engineering Chemistry Research. 2015;**54**(14):3648-3656

[166] Ashori A. Hybrid composites from waste materials. Journal of Polymers and the Environment. 2010;**18**(1):65-70

[167] Saba N, Jawaid M, Alothman OY, Paridah M. A review on dynamic mechanical properties of natural fibre reinforced polymer composites. Construction and Building Materials. 2016;**106**:149-159

[168] Thwe MM, Liao K. Effects of environmental aging on the mechanical properties of bamboo–glass fiber reinforced polymer matrix hybrid composites. Composites Part A: Applied Science and Manufacturing. 2002;**33**(1):43-52

[169] Jawaid M, Khalil HA. Cellulosic/synthetic fibre reinforced polymer hybrid composites: A review. Carbohydrate Polymers. 2011;**86**(1):1-18

[170] Jacob M, Francis B, Thomas S, Varughese K. Dynamical mechanical analysis of sisal/oil palm hybrid fiber-reinforced natural rubber composites. Polymer Composites. 2006;**27**(6):671-680

[171] Birnin-Yauri AU, Ibrahim NA, Zainuddin N, Abdan K, Then YY, Chieng BW. Enhancement of the mechanical properties and dimensional stability of oil palm empty fruit bunch-kenaf core and oil palm mesocarp-kenaf core hybrid fiber-reinforced poly (lactic acid) biocomposites by borax decahydrate modification of fibers. BioResources. 2016;**11**(2):4865-4884

[172] Fu S-Y, Lauke B, Mäder E, Yue C-Y, Hu X. Tensile properties of short-glass-fiber-and short-carbon-fiber-reinforced polypropylene composites. Composites Part A: Applied Science and Manufacturing. 2000;**31**(10):1117-1125

[173] Jacob M, Thomas S, Varughese KT. properties of sisal/oil palm hybrid fiber reinforced natural rubber composites. Composites Science and Technology. 2004;**64**(7-8):955-965

[174] Zhu Z, Ye C, Fu W, Wu H. Improvement in mechanical and thermal properties of polylactic acid biocomposites due to the addition of hybrid sisal fibers and diatomite particles. International Journal of Polymer Analysis and Characterization. 2016;**21**(5):365-377

[175] Hamid MRY, Ab Ghani MH, Ahmad S. Effect of antioxidants and fire retardants as mineral fillers on the physical and mechanical properties of high loading hybrid biocomposites reinforced with rice husks and sawdust. Industrial Crops and Products. 2012;**40**:96-102

[176] Ma P-C, Liu M-Y, Zhang H, Wang S-Q, Wang R, Wang K, et al. Enhanced electrical conductivity of nanocomposites containing hybrid fillers of carbon nanotubes and carbon

black. ACS Applied Materials & Interfaces. 2009;**1**(5):1090-1096

[177] Bao L, Zang J, Li X. Flexible Zn_2SnO_4/MnO_2 core/shell nanocable– carbon microfiber hybrid composites for high-performance supercapacitor electrodes. Nano Letters. 2011;**11**(3):1215-1220

[178] Hayase G, Kanamori K, Abe K, Yano H, Maeno A, Kaji H, et al. Polymethylsilsesquioxane–cellulose nanofiber biocomposite aerogels with high thermal insulation, bendability, and superhydrophobicity. ACS Applied Materials & Interfaces. 2014;**6**(12):9466-9471

[179] Jacob M, Varughese K, Thomas S. Dielectric characteristics of sisal–oil palm hybrid biofibre reinforced natural rubber biocomposites. Journal of Materials Science. 2006;**41**(17):5538-5547

[180] Ashik K, Sharma RS. A review on mechanical properties of natural fiber reinforced hybrid polymer composites. Journal of Minerals and Materials Characterization and Engineering. 2015;**3**(05):420

[181] Nunna S, Chandra PR, Shrivastava S, Jalan A. A review on mechanical behavior of natural fiber based hybrid composites. Journal of Reinforced Plastics and Composites. 2012;**31**(11):759-769

[182] Mishra S, Mohanty A, Drzal L, Misra M, Parija S, Nayak S, et al. Studies on mechanical performance of biofibre/glass reinforced polyester hybrid composites. Composites Science and Technology. 2003;**63**(10):1377-1385

[183] Shahzad A. Impact and fatigue properties of hemp–glass fiber hybrid biocomposites. Journal of Reinforced Plastics and Composites. 2011;**30**(16):1389-1398

[184] Velmurugan R, Manikandan V. Mechanical properties of palmyra/glass fiber hybrid composites. Composites Part A: Applied Science and Manufacturing. 2007;**38**(10):2216-2226

[185] Ahmed KS, Vijayarangan S. Tensile, flexural and interlaminar shear properties of woven jute and jute-glass fabric reinforced polyester composites. Journal of Materials Processing Technology. 2008;**207**(1-3):330-335

[186] Liu W, Drzal LT, Mohanty AK, Misra M. Influence of processing methods and fiber length on physical properties of kenaf fiber reinforced soy based biocomposites. Composites Part B: Engineering. 2007;**38**(3):352-359

[187] Benayahu D, Sharabi M, Pomeraniec L, Awad L, Haj-Ali R, Benayahu Y. Unique collagen fibers for biomedical applications. Marine Drugs. 2018;**16**(4):102

[188] Namvar F, Jawaid M, Tanir PM, Mohamad R, Azizi S, Khodavandi A, et al. Potential use of plant fibres and their composites for biomedical applications. BioResources. 2014;**9**(3):5688-5706